WOMEN PEACE & SECURITY
IN PROFESSIONAL MILITARY EDUCATION

WOMEN PEACE & SECURITY IN PROFESSIONAL MILITARY EDUCATION

EDITED BY LAUREN MACKENZIE, PhD
and LIEUTENANT COLONEL DANA PERKINS, PhD

MCUP
MARINE CORPS UNIVERSITY PRESS

Quantico, Virginia
2022

LIBRARY OF CONGRESS CATALOGING-IN-PUBLICATION DATA

Names: Mackenzie, Lauren, 1976- editor. | Perkins, Dana, 1965- editor. | Marine Corps University (U.S.). Press, issuing body.
Title: Women, peace, & security in professional military education / edited by Lauren Mackenzie and Dana Perkins.
Other titles: Women, peace, and security in professional military education
Description: Quantico, Virginia : Marine Corps University Press, 2022. | Includes bibliographical references and index. | Summary: "The Women, Peace, and Security (WPS) Agenda is a global framework and policy tool that guides national actions addressing gender inequalities and the drivers of conflict and its impact on women and girls. By fostering structural and institutional change, the WPS agenda aims to 1) prevent conflict and all forms of violence against women and girls and 2) ensure the inclusion and participation of women in peace and security decision-making processes to incorporate their specific needs in relief and recovery situations. This volume gathers together student papers from the Joint Women, Peace, and Security Academic Forum's 2021 WPS in PME Writing Award program, a best-of selection of informative and empowering work that intersects with Department of Defense equities supporting global WPS principles. Student participants in the Joint WPS Academic Forum hail from prestigious DOD academic institutions, and this monograph shows how the strategic leaders of tomorrow embrace WPS today, offering a strong indication of how WPS principles will be implemented over time and how they will influence the paradigm of peace and security and our approaches to conflict prevention and resolution"— Provided by publisher.
Identifiers: LCCN 2022024550 (print) | LCCN 2022024551 (ebook) | ISBN 9798985340365 (paperback) | ISBN 9798985340372 (epub)
Subjects: LCSH: Women and peace. | Women and human security. | Women and the military. | Women—Violence against—Prevention—International cooperation. | National security—United States. | Military education—United States. | United States—Armed Forces—Women. | United Nations. Security Council. Resolution 1325.
Classification: LCC JZ5578 .W6674 2022 (print) | LCC JZ5578 (ebook) | DDC 327.1/72082—dc23/eng/20220705 | SUDOC D 214.513:W 84
LC record available at https://lccn.loc.gov/2022024550
LC ebook record available at https://lccn.loc.gov/2022024551

DISCLAIMER

The views expressed in this publication are solely those of the authors. They do not necessarily reflect the opinion Marine Corps University, the U.S. Marine Corps, the U.S. Navy, the U.S. Army, the U.S. Air Force, or the U.S. government. The information contained in this book was accurate at the time of printing. Every effort has been made to secure copyright permission on excerpts and artworks reproduced in this volume. Please contact the editors to rectify inadvertent errors or omissions. In general, works of authorship from Marine Corps University Press (MCUP) are created by U.S. government employees as part of their official duties and are not eligible for copyright protection in the United States; however, some of MCUP's works available on a publicly accessible website may be subject to copyright or other intellectual property rights owned by non-Department of Defense (DOD) parties. Regardless of whether the works are marked with a copyright notice or other indication of non-DOD ownership or interests, any use of MCUP works may subject the user to legal liability, including liability to such non-DOD owners of intellectual property or other protectable legal interests.

The production of this book and other MCUP products is graciously supported by the Marine Corps University Foundation.

Published by
Marine Corps University Press
2044 Broadway Drive
Quantico, VA 22134
1st Printing, 2022
ISBN 979-8-9853403-6-5

THIS VOLUME IS FREELY AVAILABLE AT WWW.USMCU.EDU/MCUPRESS

CONTENTS

Foreword
    by Brigadier General Rebecca Sonkiss, USAF   IX

Preface
    by Colonel Veronica Oswald-Hrutkay, USAR   XI

Select Abbreviations and Acronyms   XIII

Introduction
    by Lieutenant Colonel Dana Perkins, PhD, USAR   XV

PART I
Implementing and Incorporating a Gender Perspective   3

Chapter 1   5
Introduction

Chapter 2   9
Closing the Capability Gap
by Lieutenant Colonel Ellen I. Coddington, USA

Chapter 3   24
A Strategic Imperative
by Commander Kristen Vechinski, USN

Chapter 4   39
A Kotter Approach for Geographic Combatant Commands
by Colonel Steven J. Siemonsma, ARNG

Chapter 5   54
Conclusion
by Colonel Douglas Winton, USA

PART II
Gender Neutrality   59

Gender-Neutral Physical Fitness Tests   61
and the Integration of Women in Combat Arms Occupations
by Second Lieutenant Elizavetta Fursova

PART III
Gender and Violence   83

Breaking a Vicious Cycle: 85
**Systemic Endorsement of Violence Against Women
in El Salvador**
by Captain Elizabeth Jane Garza-Guidara, USAF

PART IV
Professional Military Education   95

The Strategic Centrality of Women, Peace, and Security: 97
A Call to Mainstream in Professional Military Education
by Lieutenant Colonel Casey M. Grider, USAF

PART V
Vietnam   105

The Nexus of Climate Change, Migration,   107
and Human Trafficking
by Ms. Amy Patel

PART VI
Hegemonic Masculinity   115

The Effect of Hegemonic Masculinities
on the Endemic of Sexual Misconduct in the U.S. Army
by Major Sarah E. Salvo, USA

    Chapter 1   117
       Introduction

    Chapter 2   127
       Literature Review

    Chapter 3   139
       Research Methodology

    Chapter 4   142
       Analysis

    Chapter 5   161
       Conclusions and Recommendations

PART VII
Advising with Gendered Perspectives   167

Bridging the Gap toward a Gendered Perspective
in Security Force Advising
by Lieutenant Colonel Natalie Trogus, USMC

    Chapter 1   169
       Introduction

    Chapter 2   177
       Background on UNSCR 1325 Women, Peace, and Security

Chapter 3 — Advising in Wartime: A Foreign Policy Tool ... 188

Chapter 4 — Advising Afghan Security Institutions and Afghan National Defense and Security Forces ... 194

Chapter 5 — Recommendations and Conclusions ... 208

PART VIII
Implementing Women, Peace, and Security ... 213

Operationalizing Women, Peace, and Security in the Armed Sevices: Army Strategic Implementation Plan
by Major Danielle Villanueva, USA

Chapter 1 — Introduction ... 215

Chapter 2 — Literature Review ... 217

Chapter 3 — Background of WPS in the Unites States ... 221

Chapter 4 — WPS Implementation: The Australian Defence Force ... 229

Chapter 5 — U.S. Army WPS Implementation Plan: Recommended Framework ... 232

Chapter 6 — Conclusion and Research Recommendations ... 237

Epilogue ... 239
by Lauren Mackenzie, PhD

Appendices
- Appendix A. Setting the Example ... 242
- Appendix B. Advising Project Interview ... 246
- Appendix C. Integrating WPS into PME ... 249

WPS Chronology ... 253

Glossary of Key WPS Concepts and Terms ... 254

Selected Bibliography ... 257

Index ... 262

About the Editors ... 269

FOREWORD

As the deputy director of the Joint Staff J5, Counter Threats and International Cooperation, I oversee the Women, Peace, and Security (WPS) portfolio, part of the Department of Defense's (DOD) WPS program intended to promote the safety, equality, and meaningful contributions of women. From a national security perspective, advancing WPS is not just the right thing to do, it is the smart thing to do. A vast body of evidence unequivocally demonstrates that women's contributions are critical to conflict prevention and resolution, peace building and peacekeeping, disaster preparedness and recovery, and stability worldwide. We know that when women actively and meaningfully contribute to defense and security sectors, their nations are more secure and stable and more inclusive of the entire population, and that they better detect threats and vulnerabilities and better ensure the different security needs of vulnerable populations. Additionally, WPS upholds our nation's commitment to human rights and makes the United States a more credible partner of choice by supporting the development of professional, diverse, and accountable partner-nation militaries. This is why, in this era of strategic competition, it is critical that DOD integrates WPS principles to build diverse teams, incorporates a holistic and human-centered approach to security, enhances long-term stability, and increases Joint Force lethality through recruitment and retention of the most capable warriors.

To further hone our lethal force, we work to educate servicemembers on the value of WPS principles. Our Joint Staff team has developed two Joint Knowledge Online courses, available to all DOD personnel, which provide a baseline understanding of WPS and its application to key DOD mission areas: DOD Introduction to Women, Peace, and Security and DOD Women, Peace, and Security Implementation. Our professional military education (PME) institutions are currently developing Service-specific courses to ensure their members understand how WPS applies across mission areas. All WPS PME leads have all been trained in our Joint Staff Operationalizing WPS 100 (Gender Focal Point) and Operationalizing WPS 200 (Gender Advisor) courses. From conducting gender analyses for information operations campaigns and integrating gender perspectives into campaign and contingency plans to applying gender injects to disaster response wargames, our WPS PME leads are well-equipped with the knowledge and skills to apply key concepts to national security problems and to interpret their relevance to U.S. military objectives. Our Joint Staff team regularly engages with PME leads in the Joint WPS Academic Forum, where best practices and lessons learned for building curricula are shared. WPS PME leads have also been instrumental in providing feedback to the Joint Staff team developing key leader seminar materials to educate DOD senior leaders. I am proud of the work of the PME institutions, and I look forward to seeing how they drive even more action on integrating WPS principles across all courses.

I had the pleasure of reading this year's Joint WPS Academic Forum's WPS Top Paper Award-winning piece by U.S. Army major Sarah Salvo. I was impressed by Major Salvo's academic rigor. This paper highlights a key WPS principle—that advancing women's participation as a core tenet of our national security reflects DOD's commitment to diverse teams and DOD's adherence to military justice, accountability, and standards of conduct. I am encouraged to see military students addressing such important issues as sexual harassment and assault in our armed forces. The Joint Staff takes these reflections and the many recommendations and tasks from the 2021 Independent Review Commission on Sexual Assault in the Military and is striving to educate and develop a Joint Force with improved teamwork and combat capability.

Thank you to Dr. Lauren Mackenzie for developing this special monograph on WPS for the Marine Corps University Press to highlight the best papers from across the Services submitted to the Joint WPS Academic Forum's writing award program. I look forward to seeing more PME institutions spotlight WPS scholarship, which not only empowers military students but also enhances the meaningful participation of women in the defense and security sector.

Brigadier General Rebecca Sonkiss
Deputy Director, Joint Staff J5
Counter Threats and International Cooperation

PREFACE

The Joint Women, Peace, and Security (WPS) Academic Forum is the inspiration behind this special collection of student research papers that span a number of prestigious Department of Defense (DOD) academic institutions. Following the release of the 2019 *United States Strategy on Women, Peace, and Security* (WPS Strategy) and in anticipation of the DOD's *Women, Peace, and Security Strategic Framework and Implementation Plan* (WPS SFIP) publication, several senior Service colleges began collaborating to identify best practices for incorporating WPS in Joint professional military education (PME).[1] The WPS Strategy approach highlighted under its third line of effort that "the United States Government must equip and empower . . . military . . . personnel to advance the goals of this strategy . . . through an ongoing process of training, education, and professional development in partnership with specialists who can provide insight and understanding to this challenging field."[2] The issuance of the WPS SFIP further underscored this guidance. Defense Objective 1 states, "The Department of Defense exemplifies a diverse organization that allows for women's meaningful participation across the development, management, and employment of the Joint Force." Under this objective, "WPS principles are appropriately reflected in relevant [DOD] policies, plans, doctrine, education, operations, resource planning, and exercises."[3]

The first Joint WPS Academic Forum kicked off in January 2020 and the forum continues to grow in the number of participating defense institutions including academies, universities, war colleges, and regional centers. The Joint Staff J7 WPS also participates and supports the forum where needed. As the U.S. Army War College director of WPS Studies, I served as the primary facilitator and lead on behalf of the forum.

The vision of the forum is to share best practices to advance the integration of WPS concepts across PME institutions. Forum members identify opportunities to support one another and, where appropriate, synchronize efforts around common goals. The mission of the Joint WPS Academic Forum is to bring together PME institutions to build awareness and improve understanding of WPS concepts and principles to facilitate their integration into Joint and Service-specific PME programming. The forum also promotes increasing the participation of women from allied and partner nations in PME activities in support of the goals in the WPS Strategy and the WPS SFIP. The forum's lines of effort are

---

[1] *United States Strategy on Women, Peace, and Security* (Washington, DC: Office of the President, White House, 2019); and *Women, Peace, and Security Strategic Framework and Implementation Plan* (Washington, DC: U.S. Department of Defense, 2020).
[2] *United States Strategy on Women, Peace, and Security*, 11.
[3] *Women, Peace, and Security Strategic Framework and Implementation Plan*, 12, 16.

- Integrate WPS concepts and principles within PME.
- Conduct research on WPS topics to improve operationalizing concepts across the doctrine, organization, training, materiel, leadership and education, personnel, facilities, and policy (DOTMLPF-P) capability domains including capacity building within security cooperation activities.
- Advance diversity and inclusion initiatives that enable female officer, noncommissioned officer, and civilian career progression as PME faculty and staff members.

The forum meets monthly and includes discussion with various subject matter experts to advance collective goals. In August 2020, the forum conducted an inaugural two-day Joint WPS Academic Forum Workshop hosted by the U.S. Army War College. As part of these efforts, different institutions volunteered to lead a special effort in support of surging the forum initiatives. Dr. Lauren Mackenzie of Marine Corps University volunteered to lead the Best of WPS in PME Writing Award. Lieutenant Colonel Dana Perkins of Army War College, Dr. Brenda Oppermann of U.S. Naval War College, and Dr. Susan Yoshihara of Defense Security Cooperation University comprised the selection committee. The winner of the award received a Joint Staff J-5 certificate and a personalized note from me.

On behalf of the Joint WPS Academic Forum and Marine Corps University, we hope you find this special edition of the Best of WPS in PME Writing Award student research paper submissions informative and empowering, as many topics within this monograph intersect the DOD equities supporting global WPS principles.

Colonel Veronica Oswald-Hrutkay, U.S. Army Reserve
Director of WPS Studies
Office of the Provost
U.S. Army War College

SELECT ABBREVIATIONS AND ACRONYMS

ANA—Afghan National Army
ANDSF—Afghan National Defense and Security Forces
CRSV—conflict related sexual violence
CTIP—combating trafficking in persons
DOD—Department of Defense
FET—Female Engagement Team
GCC—geographic combatant command
GENAD—gender advisor
GFP—gender focal point
ISIS—Islamic State of Iraq and Syria
ISIS-K—ISIS-Khorasan
NAP—National Action Plan
NATO—North Atlantic Treaty Organization
NCO—noncommissioned officers
NDS—National Defense Strategy
NGO—nongovernmental organization
PKSOI—Peacekeeping and Stability Operations Institute
PME—professional military education
RSM—Resolute Support Mission
SFIP—Strategic Framework and Implementation Program
TAA—train, advise, assist
UNSC—UN Security Council Resolution
USAFRICOM—U.S. Africa Command
USAID—U.S. Agency for International Development
USEUCOM—U.S. European Command
USINDOPACOM—U.S. Indo-Pacific Command
USSOUTHCOM—U.S. Southern Command
WPS—Women, Peace, and Security

INTRODUCTION

The Women, Peace, and Security (WPS) agenda is a global framework and policy tool that guides national actions addressing gender inequalities and the drivers and impact of conflict on women and girls. The principles enshrined in the WPS agenda call for the prevention of conflict and for all forms of violence against women and girls in conflict and post-conflict situations; for meaningful participation of women in peace and security decision-making processes at all levels; for the protection of women and girls from all forms of sexual and gender-based violence and protection of their rights in conflict situations; and for ensuring the participation of women and meeting their specific needs in relief and recovery situations. Moreover, the WPS agenda seems to be emerging as a mechanism in the pursuit of inclusive policies and in addressing drivers of conflict and insecurity, for example, masculinity as a socially constructed norm associated with power, control, and violence. The importance of these lines of effort in fostering structural and institutional change cannot be understated.

Integrating these WPS principles across the defense enterprise is aimed at ensuring human security for all while improving operational effectiveness. How the implementation of WPS principles across the Joint Force will, in time, influence our paradigm of peace and security and our approaches to conflict prevention and resolution depends on the groundswell of strategic leaders promoting change and their education and training. This monograph shows how the strategic leaders of tomorrow embrace the WPS agenda today.

The Joint Chiefs of Staff report *Developing Today's Joint Officers for Tomorrow's Ways of War* calls for knowledge and skills of warfighting Joint leaders, senior staff officers, and strategists who can "discern the military dimensions of a challenge affecting national interest, frame the issue at the policy level, and recommend viable military options within the overarching frameworks of globally integrated operations."[1] The critical thinking and depth of knowledge of students on the intersectional and holistic nature of WPS issues in the current operational environment were evident in all papers submitted to the WPS writing competition coordinated by the Joint WPS Academic Forum. Papers were submitted by students at U.S. Air Force Air War College, Air Command and Staff College, National Defense University, U.S. Army Command and General Staff College, Army War College, and Marine Corps University; they were judged by select members of the Joint WPS Academic Forum. The topics addressed ranged from country case studies; analysis and recommendations to improve the integration of WPS into the PME/Joint PME curriculum and predeployment training across Services toward more effective decisionmaking, strategy development, and mission execution, based on lessons learned during deployment; equality and equity in the military as related to personnel readiness, physical fitness, and integration of women in combat arms occupations; the need for a culture of change based on John

---

[1] *Developing Today's Joint Officers for Tomorrow's Ways of War: The Joint Chiefs of Staff Vision and Guidance for Professional Military Education and Talent Management* (Washington, DC: Office of the Joint Chiefs of Staff, 2020), 4.

Kotter's Leading Change model and the role of leadership in combatting sexual and gender-based violence in the military; and discussion of perceived doctrinal gaps such as a putative gender in Army operations doctrinal publication and guidance. The award went to the paper by Major Sarah Salvo, of the U.S. Army Command and General Staff College, "The Effect of Hegemonic Masculinities on the Endemic of Sexual Misconduct in the U.S. Army," which provides critical and nuanced insights on the aspects of organizational culture that undermine trust and contribute to sexual abuses.

With reference to the PME outcome envisioned by the Joint Chiefs of Staff, all students demonstrated critical and creative thinking skills and effective communication skills to operationalize WPS and support the development and implementation of WPS strategies and complex operations. Continuing to leverage and integrate WPS in the PME/Joint PME curricula and the doctrine, organization, training, materiel, leadership and education, personnel, facilities, and policy (DOTMLPF-P) spectrum has the potential to significantly strengthen the human capital and the elements of national power in great power competition, crisis, and conflict in order to achieve the objectives of national security.

Lieutenant Colonel Dana Perkins, PhD
U.S. Army Reserve, MS
Director of Women, Peace, and Security Studies
Office of the Provost, U.S. Army War College

WOMEN PEACE & SECURITY
IN PROFESSIONAL MILITARY EDUCATION

PART 1

Implementing and Incorporating a Gender Perspective

Chapter 1
Introduction

On 31 October 2000, the United Nations (UN) drafted and approved Security Council Resolution (UNSCR) 1325 on Women, Peace, and Security (WPS). The adoption of this resolution was a landmark event as it recognized "the disproportionate impact of armed conflict on women" in addition to emphasizing the "importance of women's equal and full participation in the prevention and resolution of conflicts, peacekeeping, and peacebuilding."[1] The United States did not develop a framework through which the WPS could be implemented across the Department of Defense (DOD) for two decades after the passage of UNSCR 1325. The recent William M. (Mac) Thornberry National Defense Authorization Act for Fiscal Year 2021 (NDAA) requires the DOD to develop a strategic framework and implementation plan and identifies the secretary of defense as the lead agent for its employment. Included in this language is the requirement to address the following three objectives: implementing the 2020 *Women, Peace, and Security Strategic Framework and Implementation Plan* (WPS SFIP); establishing policies and programs that reinforce the SFIP objectives and equities; and adequately training personnel to advance the WPS agenda. Further, the NDAA specifies this action is to be completed by the end of September 2025.[2]

The WPS SFIP is broad, and the authors herein elected to initially focus on ways in which a gender perspective can serve as a capability across the Joint Force. Each presents definitions of the problem in separate chapters. When combined, they address many of the challenges associated with advancing the WPS agenda. With greater than 50 percent of the American population being female, it was essential that we address what is meant by a gender perspective as the foundational step for research, before examining the other facets of WPS-related research.[3]

It is important to note that implementing a gender perspective is not limited to one gender, nor is it about women's rights, feminism, or any other women's issue. The value of a gender perspective lies in its critical examination of behaviors, attitudes, and beliefs. The DOD has worked diligently to address the patriarchal structure of the past; however, much work is left, requiring a culture change. The DOD must build awareness and address behaviors and biases, which left unchecked may further delay achieving many of the WPS objectives.[4]

Army lieutenant colonel Ellen I. Coddington, authors the first piece in this series, "Closing the Capability Gap." In this initial chapter, she introduces the gap as it exists

---

[1] Anne A. Witkowsky, "Integrating Gender Perspectives within the Department of Defense," *Prism* 6, no. 1 (March 2016): 35.
[2] William M. (Mac) Thornberry National Defense Authorization Act for Fiscal Year 2021, Pub. L. No. 116–283, (2020).
[3] Female persons, percent, "Quick Facts, United States," U.S. Census Bureau, accessed 11 February 2022.
[4] "The Pentagon Has a Plan to Include More Women in National Security. Here's What That Means—and Why It Matters," *Task and Purpose*, 10 July 2020.

in the organization and addresses the first of three objectives in the SFIP: a diverse and inclusive fighting force.[5] This objective cannot be met within a culture where divisiveness, fear, and mistrust are common. Coddington further examines how three of the equities discussed in the SFIP—diversity and inclusion, sexual harassment assault and prevention, and gender-based violence—negatively impact military culture and cohesiveness. For this piece, she examined parts of the final report of the Fort Hood Independent Review Committee (FHIRC), formed as a result of the death of U.S. Army Specialist Jessica Guillén in April 2020. Although the FHIRC focused its investigation on Fort Hood, its findings and recommendations will ultimately benefit the U.S. Army and most likely the DOD writ large.[6]

Significant strides have indeed been made over the last few decades in advancing the role of women within the various U.S. defense agencies. Women have been authorized and accepted on a volunteer basis to pursue combat military occupational specialties and units, and they have been admitted to and graduated from specialized schools, such as Ranger School. Additionally, there have been uniform changes and relaxing of grooming standards, yet there is still much work to be done. Formations across DOD are becoming more diverse both in race and gender, and even this continues to evolve with the further acceptance of gay, lesbian, and transgender servicemembers.

Diversity and inclusion are the most logical first step that must be seriously considered before the other equities can ever be achieved. The absence of this equity fosters a permissive and toxic environment where other questionable behaviors can fester and prevail. Thus, sexual harassment, often referred to as *gateway behavior*, also becomes an acceptable behavior, setting the environment to become capable of escalating to sexual assault and gender-based violence.

The second offering in this series, USN Commander Kristen Vechinski's "A Strategic Imperative," focuses on applying WPS in military operations to achieve effectiveness in the security environment. Here, Vechinski suggests that delaying the WPS agenda—specifically incorporating gender perspective—puts the United States and the Joint Force at a disadvantage in the security environment. Anecdotal evidence intimates that incorporating gender perspective as a capability improves operational effectiveness when applied to the planning and execution of information gathering, increasing credibility and legitimacy, and enhanced force protection.[7]

When considering the current security environment, the Joint Force must demonstrate proficiency across the competition continuum, denying and deterring adversaries any gains or advantages. Vechinski examines how female engagement teams

---

[5] *Women, Peace, and Security Strategic Framework and Implementation Plan* (Washington, DC: U.S. Department of Defense, 2020), 7.
[6] *Report of the Fort Hood Independent Review Committee* (Washington, DC: Office of the Secretary, U.S. Army, 2020), 121–22.
[7] Kristen D. Vechinski, "A Capability for Improved Military Effectiveness" (strategic research paper, U.S. Army War College, 2021).

employed in Afghanistan facilitated gains during counterinsurgency operations using gender-focused strategies in the human domain.[8]

The DOD must move to institutionalize WPS principles as a fundamental part of full-spectrum military operations. As demonstrated in the Afghanistan case study, the incorporation of gender perspectives into operations will build capacity within the Joint Force, improving efficiency and increasing lethality in support of national defense objectives.

The third and final contribution to this work, ARNG Colonel Steven J. Siemonsma's "A Kotter Approach for Geographic Combatant Commands," directs attention to the global combatant commands as a focal point to begin the culture changes necessary to operationalize gender perspective and introduce WPS principles globally. At present, the geographic combatant commands (GCCs) still exhibit many of the characteristics associated with the patriarchal organization, even as the United States criticizes other nations abroad for continued gender-based inequalities within their organizations and cultures.[9] The United States is a global leader, and, in this position, must set the standard. When the United States shows up for the fight, it must be willing to lead. Simply put, the United States cannot set expectations for everyone else and then do something different. When situated abroad, the GCCs are in an ideal position and location to institute this much-needed culture change as a forward presence, lending credibility to change efforts and the facilitation of democratic values abroad.

As a result of his research, Siemonsma recommends that the GCCs employ the eight-step process described in Dr. John Kotter's book *Leading Change* (1988) to guide the cultural changes required to implement a gender perspective at the organizational level successfully. This process will enable the GCCs to incorporate a systematic approach in developing effective programs and strategies, enabling leadership to facilitate opportunities and exploit the advantages of gender perspective and WPS principles at home and abroad.[10]

As the reader navigates through these three chapters, they will note several common themes. To successfully implement WPS strategies, DOD must first address its culture across the entirety of its battlespace. High-functioning organizations are mission-focused, trained and ready, and foster a culture and environment rich in diversity, inclusion, dignity, respect, and excellence. To ensure credibility with U.S. partners and allies, this must first be done at home, within the U.S. military's ranks at both the highest and lowest levels. The volatility of the global security environment demands that the implementation of WPS principles be taken seriously. The DOD must capitalize on the opportunities and advantages of incorporating the gender

---

[8] Vechinski, "A Capability for Improved Military Effectiveness."
[9] Steven J. Siemonsma, "A Kotter Approach for Geographic Combatant Commands" (strategic research paper, U.S. Army War College, 2021).
[10] Siemonsma, "A Kotter Approach for Geographic Combatant Commands."

perspective as a capability. Senior leaders across DOD must become engaged in deliberating effective policies that implement the WPS SFIP through doctrinal changes, training, advocacy, and culture change that will ensure a diverse, inclusive, and lethal fighting force.

Chapter 2
Closing the Capability Gap
by Lieutenant Colonel Ellen I. Coddington, USA*

Close to half of the world's population is female. Yet, while human rights for women has been a subject often noted in numerous documents, it usually only gains periodic attention during a conflict or humanitarian crisis. This passive stance started to change in the 1990s when UN member states began to reconsider international norms.[11] The Fourth World Conference on Women hosted in Beijing, China, in 1995 was a turning point; it was one of the first events to trigger the establishment of an international agenda focused on "equality, development, and peace for women."[12] Regrettably, it would be another five years, on 31 October 2000, before the United Nations would draft and approve UNSCR 1325, Women, Peace, and Security. The Security Council's adoption of this resolution was a landmark event as it recognized "the disproportionate impact of armed conflict on women" in addition to emphasizing the "importance of women's equal and full participation in the prevention and resolution of conflicts, peacekeeping, and peacebuilding."[13] It would take two more decades after the Security Council's action for the DOD to develop a framework and plan for implementing the WPS agenda.

Since the Security Council first acted, DOD has missed critical opportunities to implement the WPS principles, which it has since articulated in its *Women, Peace, and Security Strategic Framework and Implementation Plan*. Moreover, the armed Services perpetuate an environment steeped in controversy related to various forms of sexual misconduct, harassment, and abuse. Further, DOD struggles with diversity and inclusion policies that are focused, relevant, defined, and enduring. These issues are intrinsically linked, and one cannot be adequately addressed without awareness, understanding, and attention to the others. Nonetheless, mitigation of misconduct, harassment, and abuse can occur if DOD promulgates policies that address these problems. By focusing on the three equities indicated in the DOD plan, specifically diversity and inclusion, gender-based violence prevention, and sexual harassment and assault prevention, the department will promote successful implementation of its WPS objectives by fostering a gender perspective that raises awareness, changes its institutional culture, and decreases the capability gap that exists in the current Joint Force.

---

* The views expressed in this chapter are solely those of the author. They do not necessarily reflect the opinion of Marine Corps University, the U.S. Marine Corps, the U.S. Navy, the U.S. Army, U.S. Army War College, the U.S. Air Force, or the U.S. government.
[11] *Women's Rights are Human Rights* (New York and Geneva: Office of the United Nations, High Commissioner for Human Rights, 2014), 11–13.
[12] Anne A. Witkowsky, "Integrating Gender Perspectives within the Department of Defense," *Prism* 6, no. 1 (March 2016): 35.
[13] Witkowsky, "Integrating Gender Perspectives within the Department of Defense," 35.

Background

UNSCR 1325 recognized "the urgent need to mainstream gender perspective into peacekeeping operations" in addition to identifying and reaffirming the significance of the role that women and girls play pre-, during, and post-conflict as it relates to peaceful resolution and stability.[14] UNSCR 1325 established the following four principles that support the WPS model: 1) ensuring women's participation at all levels of decision-making in the peace and security sphere; 2) protecting the rights of women and girls; 3) incorporating a gender perspective into conflict prevention initiatives; and 4) making certain that gender considerations are integrated into relief and recovery efforts.[15]

It was almost 11 more years before the United States took its first steps toward addressing this crisis at the national level in the form of an executive order and a plan. The *United States National Action Plan on Women, Peace, and Security* was published in December 2011 in response to President Barack Obama's Executive Order 13,595, Instituting a National Action Plan on Women, Peace, and Security. The plan highlighted the gender integration and perspective goals identified in the 2010 *National Security Strategy* and the 2011 *Quadrennial Diplomacy and Development Review*.[16] Yet, it would be another six years before President Donald J. Trump signed into law the Women, Peace, and Security Act of 2017, the first statute of its kind, establishing policy, defining strategy, specifying training requirements, and identifying DOD, the Department of State, the U.S. Agency for International Development, and the Department of Homeland Security as the lead agencies.[17] It would take another two years before the *United States Strategy on Women, Peace, and Security* was published in June 2019.[18] It would be still another year before DOD published its *Women, Peace, and Security Strategic Framework and Implementation Plan* (WPS SFIP) in June 2020.

Defining Terms

Understanding the terms used in the WPS SFIP is critical to comprehending the issues it addresses and attaining its objectives. There are seven terms of importance for this paper. The first term is *gender perspective*. The DOD defines it as "an analytic view that examines how being treated as a man or a woman in society shapes a person's

---

[14] UN Security Council, Resolution 1325, Women, Peace and Security, S/RES/1325 (31 October 2000), hereafter UNSCR 1325.
[15] Veronica Hrutkay, "Women, Peace, and Security (WPS) National Strategy Crosswalk" (briefing presentation, U.S. Army War College, 2021); and UNSCR 1325.
[16] *United States National Action Plan on Women, Peace, and Security* (Washington, DC: Office of the President, White House, 2011), 1.
[17] Women, Peace, and Security Act of 2017, Pub. L. No. 115–68 (2017).
[18] "U.S. Strategy on Women, Peace, and Security," Department of Homeland Security, 4 October 2021.

needs, interests, control of resources, and security."[19]

Since this paper deals with equities, a definition of this term is warranted. *Equity* is "a situation in which everyone is treated fairly and equally."[20] DOD has also defined the three equities discussed in this paper. *Diversity* is "all the differing characteristics and attributes of individuals from varying demographics that are consistent with the DOD's core values, integral to overall readiness and mission accomplishment, and reflective of the nation we serve."[21] *Inclusion* is "a set of behaviors (culture) that encourages service members and civilian employees to feel valued for unique qualities and to experience a sense of belonging."[22]

Gender-based violence is "an umbrella term for any harmful threat or act directed at an individual or group based on actual or perceived biological sex, gender identity and/or expression, sexual orientation, and/or lack of adherence to varying socially constructed norms around masculinity and femininity. It is rooted in structural gender inequalities, patriarchy, and power imbalances." This form of violence is "typically characterized by the use or threat of physical, psychological, sexual, economic, legal, political, social, and other forms of control and abuse." It also "impacts individuals across the life course and has direct and indirect costs to families, communities, economies, global public health, and development."[23]

Sexual harassment and sexual assault are constituted as the third equity. DOD defines *sexual assault* as the "intentional sexual contact characterized by use of force, threats, intimidation, or abuse of authority or when the victim does not or cannot consent."[24] In contrast, *sexual harassment* is defined as "conduct that involves unwanted sexual advances, requests for sexual favors, and deliberate or repeated offensive comments or gestures of a sexual nature" by Section 1561 of Title 10, United States Code.[25]

Aligning WPS with National Strategies

The WPS SFIP identifies three WPS defense objectives. These objectives seek to align the SFIP with the 2017 *National Security Strategy of the United States of America* (NSS) and the 2018 *National Defense Strategy of the United States of America* (NDS). The first is that DOD "exemplifies a diverse organization that allows for women's meaningful participation across the development, management, and employment of

---

[19] DOD Introduction to Women, Peace, and Security course and DOD Women, Peace, and Security Implementation course, both Joint Knowledge Online, hereafter JKO WPS training courses.
[20] Women, Peace and Security glossary resource for JKO WPS training courses.
[21] *Department of Defense Board on Diversity and Inclusion Report: Recommendations to Improve Racial and Ethnic Diversity and Inclusion in the U.S. Military* (Washington, DC: U.S. Department of Defense, 2020), 3.
[22] *Department of Defense Board on Diversity and Inclusion Report*, 3.
[23] Women, Peace and Security glossary resource for JKO WPS training courses.
[24] "Sexual Assault Prevention and Response," Defense Intelligence Agency, accessed 2 November 2021.
[25] *Sexual Harassment Assessment*, Appendix F, in *Department of Defense Annual Report on Sexual Assault in the Military, Fiscal Year 2019* (Washington DC: U.S. Department of Defense, 2020), 1–2.

the Joint Force." The second is that "women in partner nations meaningfully participate and serve at all ranks and in all occupations in defense and security sectors." And third, "partner nation defense and security sectors ensure women and girls are safe and secure and that their human rights are protected, especially during conflict and crisis."[26] The relationship between the SFIP and its WPS principles and these two national strategies is clear.

The NSS identifies several priorities under pillar four, "Advance American Influence," that exemplify the nation's fundamental values and are related to WPS; notably, this includes "support the dignity of individuals" and "empower women and youth."[27] The same is true for the NDS when it discusses "build[ing] a more lethal force." This approach extends beyond the modernization of equipment, force structure and employment, and weapons systems. Included here is the people factor, including recruiting and retention, talent management, and professional education for both the military and civilian workforce.[28] Cultivating a thriving workforce requires a culture and environment infused in diversity, inclusiveness, trust, respect, and safety.

Aligning the WPS SFIP and the national strategies requires more than expressing such notions in a document. The three equities that this paper examines must be tied to the three defense objectives. The equities must be addressed through two actions: 1) leaders must emphasize them, 2) and awareness programs must be developed and implemented before the first objective can be achieved. It must be satisfied utilizing a top-down approach with senior departmental leaders underscoring the equities' importance to the first objective.

The WPS SFIP indicates that defense objective one is a diverse and inclusive fighting force.[29] This objective cannot be met in the current military culture in which divisiveness, fear, and mistrust are common. Accomplishing this objective will take a comprehensive approach to combat and mitigate concerns associated with diversity and inclusion, gender-based violence, sexual harassment and assault, and other behaviors that compromise the force's cohesiveness.

Challenges to Implementing the WPS Plan and Framework

The DOD has established three objectives that support the 2020 *Women, Peace, and Security Strategic Framework and Implementation Plan*. The equity of diversity and inclusion is identified as a part of the first objective: "DoD exemplifies a diverse

---

[26] *Women, Peace, and Security Strategic Framework and Implementation Plan* (Washington, DC: U.S. Department of Defense, 2020), 7.
[27] *National Security Strategy of the United States of America* (Washington, DC: Office of the President, White House, 2017), 42.
[28] *Summary of the 2018 National Defense Strategy of the United States of America: Sharpening the American Military's Competitive Edge* (Washington DC: U.S. Department of Defense, 2018), 5–8.
[29] *Women, Peace, and Security Strategic Framework and Implementation Plan* (Washington, DC: Department of Defense, 2020), 7.

organization that allows for women's meaningful participation across the development, management, and employment of the Joint Force" and is further clarified in the document's Intermediate Defense Objective 1.1: "DoD recruitment, employment, development, retention, and promotion efforts are informed by WPS initiatives, to ensure a diverse and inclusive fighting force."[30]

The Defense Advisory Committee on Women in the Services (DACOWITS) reported in 2019 that women comprised approximately 17 percent of the U.S. military's active-duty force. Moreover, statistics demonstrate that diversity among the female population is considerably higher than the male population, with racial and ethnic diversity at 40 percent and 18 percent, respectively, for females, compared to 25 percent and 16 percent for males.[31]

All the military Services reported increases in minority populations in the decade between 2008 and 2018, with an overall increase of 2 percent for all women, a 7 percent increase in Hispanic women, and a 1 percent increase in racial minorities. For example, the Army reports that in 2018, one out of every two females is a member of a minority group. While the numbers for minority-group members are significant among enlisted women (65 percent), the overall numbers in the officer corps for minority populations is 32 percent. Additionally, the number of minority women represented in the Army's general officer population, as an example, has decreased in the last decade.[32]

Despite improvements in overall numbers across the active-duty force, women continue to experience significant challenges compared to their male counterparts when getting married and starting a family, often choosing to leave the military. Further, despite efforts to accommodate needs specific to this population, women in general still feel that current DOD policies "do not support [their] roles as mothers," forcing them to choose between motherhood and the military.[33] Recognizing the need to address this discriminatory practice prompted former secretary of defense Mark Esper to direct updates to the department's equal opportunity policies to include pregnancy-based discrimination in his July 2020 memorandum to civilian and military leaders, "Immediate Actions to Address Diversity, Inclusion, and Equal Opportunity in the Military Services."[34] Further, in his June 2020 message, "Actions for Improving Diversity and Inclusion in the Department of Defense," Esper had recognized that an institution that seeks to "embrace diversity and change" is also an enterprise that

---

[30] *Women, Peace, and Security Strategic Framework and Implementation Plan*, 7, 16.
[31] Laura M. Massey, "Concept Plan: Gender Equity Working Group" (briefing presentation, U.S. Army Equity and Inclusion Agency, 2020), slide 6; and *Defense Advisory Committee on Women in the Services 2019 Annual Report* (Alexandria, VA: Defense Advisory Committee on Women in the Services, 2019), 23.
[32] Massey, "Concept Plan," slide 6.
[33] Jennifer Rea, "Unique Challenges Faced by the Powerful Women who Serve in the United States Military," Military Families Learning Network, 14 March 2020.
[34] Mark T. Esper, DOD memo, "Immediate Actions to Address Diversity, Inclusion, and Equal Opportunity in the Military Services," 14 July 2020, hereafter Esper July 2020 memo.

will have to undergo profound organizational and cultural changes.[35] These cultural changes are themselves signs of a willingness to respect differences and not merely symbolic gestures. Instead, cultural changes should underscore the criticality of morale in building exceptional organizations and especially a "fighting force." As an example, it has only been in recent years that modifications in the combat uniform were made to accommodate most women's bodies; nonetheless, most equipment continues to be fashioned with the male physique in mind.[36]

For instance, recent policy changes outlined to *Wear and Appearance of Army Uniforms and Insignia,* Army Regulation 670-1, address several challenges associated with women, particularly those associated with minority women: lifting restrictions on hairstyles and relaxing the standards on the length of hair will circumvent severe hair loss for women.[37] Fashion aside, these revised standards offer more practical and more comfortable solutions for most women. As far as uniform standards (minus dress uniforms) are concerned, the Army has long practiced a gender-neutral standard. The addition of lipstick, nail polish, and earrings are welcome but do little to advance the diversity and inclusion agenda. Instead, the Department of the Army's December 2020 report on the Fort Hood, Texas, command climate and culture underscores the difficulties inherent in promoting cultural change to achieve the aforementioned defense objective.[38]

The Army's independent review conducted at Fort Hood and the surrounding military community following the murder of Specialist Vanessa Guillén demonstrated that there is still a lot of work to do to make notable improvements related to the diversity and inclusion equity. Statistically, Fort Hood fell short in diversity management, inclusion, sex discrimination, and racial discrimination compared to other Army installations between 2014 and 2019. In the surveys conducted for the report, 54 percent of survey participants indicated that there were indeed concerns related to how women and minorities were treated in the Army, with 44 percent demonstrating that the Army still had not done enough to level the playing field in matters relating to the promotion of women and minorities.[39]

Although survey participants were not required to provide remarks as a part of their response, some of the offered feedback was concerning. For example: "The contributions of female Soldiers in this command is still [sic] not appreciated as much as those of males—there is a definite 'boys club' among the staff and commanders."[40] Further remarks such as: "Females in this unit are not respected at all. We are often

---

[35] Mark T. Esper, "Message to the Force on DOD Diversity and Inclusiveness" (speech, U.S. Department of Defense, 18 June 2020).
[36] Rea, "Unique Challenges Faced by the Powerful Women who Serve in the United States Military."
[37] John E. Whitley, U.S. Army memo, "Appearance and Grooming Policies for the United States Army," 24 February 2021.
[38] *Report of the Fort Hood Independent Review Committee* (Washington, DC: Office of the Secretary, U.S. Army, 2020).
[39] *Report of the Fort Hood Independent Review Committee,* 121.
[40] *Report of the Fort Hood Independent Review Committee,* 121.

taunted, teased and ridiculed for going to seek medical help for sickness injuries or other female health issues. We are often seen and verbally told that we are weaker than the males and that we should not be amongst males in a combat MOS, because we are not fit."[41] Finally, "I wake up now regretting I joined the military as a young minority female and do not feel as though I fit in with my current company. I feel like an outcast often, I come to work and just sit here and talk to no one. . . . I want to be somebody and I want to be utilized, but instead I am left to defend [sic] for myself at Fort Hood with no voice."[42] While these remarks are profoundly concerning and are specific to the Fort Hood report, they are indicators of prevalent thinking within the military Services. The Fort Hood Independent Review Committee(FHIRC) certainly took note of such viewpoints and their ramifications as the committee members concluded that the "findings and recommendations" were not only "intended to benefit Fort Hood," but "the entire Army."[43] Thus, Fort Hood's problems represented a microcosm of the challenges confronting the entire Army enterprise and not a single installation and are likely representative of the challenges the other services confront.

The previous secretary of defense's policy changes and the FHIRC's recommendations are a good start but may not yet go far enough. As an example, while current policy guidance requires the removal of official photographs and demographic information from all promotion board files, this process will not be entirely neutral until personnel files are viewed strictly by DOD identification number.[44] This approach means that all demographic and defining characteristics are removed, enabling the various selection processes to be executed solely on merit. Such policy changes have ramifications well beyond promotion processes and affect how policies concerning diversity and inclusion are defined and implemented.

Gender perspective provides insight into the challenges ahead. The gender perspective is not about women, women's rights, or feminism. Even as the department plans to revise policies emphasizing and supporting the equities of diversity and inclusion, the cultural environment continues to develop. The dynamics of today's force have evolved. It is no longer binary, and the gender perspective dynamic also includes servicemembers who are gay, lesbian, and transgender. Leaders at all levels must understand how these changing dynamics will influence efforts to achieve diversity and inclusion objectives within the Services. Recognition of this change is already occurring, as the president has indicated. On 25 January 2021, President Joseph R. Biden signed Executive Order 14,004, Enabling all Qualified Americans to Serve their Country in Uniform. This executive order emphasizes that allowing transgender servicemembers to serve openly supports the national core values, strengthening the

---

[41] *Report of the Fort Hood Independent Review Committee*, 122.
[42] *Report of the Fort Hood Independent Review Committee*, 122.
[43] Army News Service, "Army Secretary Releases Results of Fort Hood Review," press release, 8 December 2020.
[44] Esper July 2020 memo.

United States and how other countries perceive it—stressing explicitly "that a more inclusive force is a more effective force."[45]

The DOD has made worthy efforts to foster a diverse and inclusive culture over the last decade. However, leaders at all levels must continue to emphasize and support equity across the force. As the Fort Hood findings indicate, fostering an environment dismissive of racial, gender, and behavioral issues poses significant challenges in protecting at-risk populations within the Services. The potential to create a culture of intolerance may be further exacerbated by a leadership team that did not appear to acknowledge the risk associated with the high crime in and around Fort Hood.[46]

Further, the lack of awareness, inconsistent application of policy and sanctions, and inadequate accountability in applying diversity and inclusion programs and the failure to reduce or mitigate instances of crime imply a permissive environment that further increases potential victimization of the local population. These factors are associated with sexual harassment and sexual assault, (i.e., incidents of gender-based violence).

Linking Sexual Harassment and Assault Prevention

The DOD must be successful in establishing an effective program focused on diversity and inclusion. Diversity and inclusion will be the crucial first step in creating an environment rich in dignity and respect, two factors that are critically important in launching an equally effective sexual assault prevention and response program. Achieving proficiency through this program supports the prevention and protection principles cited as a part of the WPS SFIP.[47] With this in mind, DOD has acknowledged that it is not uncommon that sexual harassment and assault result from progressive behaviors related to "sexually harassing and discriminatory language and behaviors."[48]

The 2005 National Defense Authorization Act (Section 577, Public Law 108-375) directed the secretary of defense to develop a policy concerning the prevention of and response to sexual assault of servicemembers. *Department of Defense Directive 6495.01, Sexual Assault Prevention and Response (SAPR) Program*, provides the department-wide policy on sexual assault prevention and response to eliminate incidents of sexual harassment and assault in the military.[49] The SAPR programs that Services have instituted use a framework focused on training, awareness, advocacy, reporting mechanisms, and personal and leadership accountability.[50] However, 15

---

[45] The White House, Briefing Room, "Fact Sheet: President Biden Signs Executive Order Enabling All Qualified Americans to Serve Their Country in Uniform," news release, 25 January 2021.
[46] Report of the Fort Hood Independent Review Committee, 90.
[47] Women, Peace, and Security Strategic Framework and Implementation Plan, 9.
[48] Carl Andrew Castro et al., "Sexual Assault in the Military," *Current Psychiatry Report* 17, no. 7 (July 2015): 9, https://doi.org/10.1007/s11920-015-0596-7.
[49] Report of the Fort Hood Independent Review Committee, 11; and *Department of Defense Directive 6495.01, Sexual Assault Prevention and Response (SAPR) Program* (Washington, DC: Department of Defense, 23 January 2012).
[50] Report of the Fort Hood Independent Review Committee, 11–12.

years after its establishment, DOD's program continues to fall under intense scrutiny at the government's top echelons. At the direction of President Biden, Secretary of Defense Lloyd Austin instructed the combatant commanders to provide feedback on "the best plans and practices" utilized in combating the persistence of sexual harassment and assault incidents across the force.[51] While Secretary Austin acknowledged that the department had undertaken numerous steps to combat the issues of sexual assault, a lot of work remains. In his memorandum to the combatant commanders, Secretary Austin emphasized that servicemembers cannot defend the United States, the department's primary mission, "if we also have to battle enemies within the ranks."[52]

Today, more than 210,000 women are serving on active duty across the Joint Force. Despite the highest number reported since World War II, this population experiences various forms of gender discrimination, sexual harassment, and sexual assault.[53] Notwithstanding the persistent discriminatory behavior experienced by many women discussed previously, sexual assault is a significant reason why numerous women choose to leave the military.[54]

The 2019 *Department of Defense Annual Report on Sexual Assault* in the military reports that sexual assault cases increased during the fiscal year 2018 by greater than 3 percent from the previous reporting period; 7,825 incidents were reported across the force.[55] Despite several years of awareness programs, annual training requirements, and emphasis at the highest level, it is overwhelming and concerning to note that during the 2018 reporting period, the number of restricted reports filed increased more than 17 percent.[56]

This increase is disheartening given that while restricted reports ensure the victim is provided care and services, there is little notification and investigation by the chain of command. Incredibly, the perpetrator is never held accountable. Despite the increase in reported sexual assault cases, the 7 percent conviction rate for these crimes remained comparable to previous reporting cycles.[57] Even as President Biden has ordered the Army's Sexual Harassment/Assault Response and Prevention (SHARP) program's review, it is still undecided if the Biden administration will remove the trial and disciplinary process from the defendant's military chain of command and instead place it under the supervision of the civilian courts.[58]

The incident at Fort Hood was a tragedy in the wake of efforts to improve the life, health, equality, and wellbeing of servicemembers—more especially in view

---

[51] Kevin Barron, "Biden Orders New Review of Sexual Assault Policies in the Military," Defense One, 25 January 2021.
[52] Barron, "Biden Orders New Review of Sexual Assault Policies in the Military."
[53] Rea, "Unique Challenges Faced by the Powerful Women Who Serve in the United States Military."
[54] Castro et al., "Sexual Assault in the Military," 2.
[55] *Department of Defense Annual Report on Sexual Assault in the Military, Fiscal Year 2019*, 6.
[56] Jennifer Steinhauer, "A #MeToo Moment Emerges for Military Women after Soldier's Killing," New York Times, 11 July 2020.
[57] Steinhauer, "A #MeToo Moment Emerges for Military Women after Soldier's Killing."
[58] Barron, "Biden Orders New Review of Sexual Assault Policies in the Military."

of the DOD policies designed to enhance the equities described in the 2020 WPS SFIP. The circumstances surrounding the disappearance and murder of Specialist Vanessa Guillén at Fort Hood, Texas, on 22 April 2020, were the culmination of a toxic environment that permitted sexual harassment and assault to transpire without accountability. Her murder, in addition to other incidents resulting in the deaths or disappearances of other soldiers, were identified as factors that resulted from the significant deficiencies in the climate at Fort Hood, disinterest on the part of Fort Hood leadership, and failure to mitigate the risks to the soldier population related to criminal activity in the communities surrounding Fort Hood. The Fort Hood community had significantly higher numbers of "violent sex crimes and other sex crimes, violent felonies, assault and battery, drug offenses, drunk and disorderly, larceny and other misdemeanors, desertions and AWOL" than other U.S. Army Forces Command installations from 2016 to 2020.[59] Most aggravating was the prevalence of sexual harassment and assault within the Fort Hood community, with Fort Hood reporting the most significant number of sexual assaults among all Army installations from 2013 to 2016, demonstrating a persistent and permissive environment lacking leadership and resources.[60] The report further discussed a culture in which the soldiers and civilians at Fort Hood lacked trust in the leadership when reporting sexual harassment and assault, gender-based violence, and diversity and inclusion complaints and concerns.[61] It is incredibly discouraging that despite years of awareness programs and annual training requirements, the installation reports the highest number of on-post sexual assaults.[62]

As discussed earlier, prevention and protection are the two most prevalent WPS principles when talking about sexual harassment and assault. The Fort Hood leaders' failure to acknowledge and mitigate the known risk at the installation meant that they did not protect the Army's most vulnerable population: enlisted female soldiers, who are most likely to be the assault victims of male peer or near-peer acquaintances.[63] This betrayal is further intensified by the lackluster attention that unit leadership paid to the components of the SHARP Program, deeming it a "perfunctory" rather than a "priority" task that focused on the team-centric safety, morale, dignity, and respect that defines the family aspect of military service.[64] This pervasive and persistent behavior compromises the readiness, trust, and cohesion of units that will further compromise the force's capabilities in meeting the objectives specified in DOD's WPS SFIP.

Escalating to Gender-Based Violence

---

[59] *Report of the Fort Hood Independent Review Committee*, 96.
[60] *Report of the Fort Hood Independent Review Committee*, 98.
[61] *Report of the Fort Hood Independent Review Committee*, 126.
[62] *Report of the Fort Hood Independent Review Committee*, 70.
[63] *Report of the Fort Hood Independent Review Committee*, 19–21.
[64] *Report of the Fort Hood Independent Review Committee*, 18.

Gender-based violence must be addressed within military formations before gender perspective can be institutionalized. The investigation into the death of Specialist Guillén exposed severe problems with the culture and environment at Fort Hood. The FHIRC uncovered evidence of an environment that tolerated sexual harassment and assault. However, the incident at Fort Hood revealed the prevalence of gender-based violence throughout the force.[65] The findings indicated that the risk of violent sex crimes at Fort Hood "were known or should have been known," and that the numbers reported from the investigation "were the highest, the most cases for sexual assault and harassment, and murders for our entire formation of the U.S. Army."[66] Guillén was just one of many victims at the installation, and her experience and, ultimately, her death, were symptomatic of a more significant, systemic problem. A considerable number of soldiers stationed at Fort Hood, male and female, reported a culture where incidents of sexual harassment, bullying, and worse, were tolerated.[67]

Gender-based violence, which encompasses both sexual harassment and assault, is widespread throughout the U.S. military and society at large.[68] It affects "families, communities, economies, global public health, and development."[69] Moreover, exercising power through the use of gender-based violence includes a wide range of activities to achieve its ends: "physical, psychological, sexual, economic, legal, political, social and other forms of control and abuse" are commonplace, and while it is most commonly directed towards women and girls, gender-based violence with men and boys as victims should not be overlooked.[70]

As discussed earlier, approximately 17 percent of the military's active-duty population is female.[71] The U.S. Department of Veterans Affairs reports that one in three female veterans experienced some form of sexual assault or harassment during their military service, with the victim being 28 percent more likely to leave or separate from service prematurely "citing sexual assault as a key factor."[72] There are concerns that the number of sexual harassment and assault incidents and incidents related to gender-based violence goes underreported during military service by many accounts. It is assumed these incidents go unreported because of individual concerns connected to fear of retaliation, adverse consequences on one's career, or social stigma and discrimination by peers and superiors. Furthermore, the effects of gender-

---

[65] Melissa E. Dichter, Gala True, and Glenna Tinney, "A Call to End Gender-Based Violence in the Military," *Military Times*, 11 February 2021.
[66] Manny Fernandez, "A Year of Heartbreak and Bloodshed at Fort Hood," *New York Times*, 9 September 2020.
[67] Fernandez, "A Year of Heartbreak and Bloodshed."
[68] Dichter, True, and Tinney, "A Call to End Gender-Based Violence in the Military."
[69] Beth Lape, "Gender-Based Violence" (briefing presentation, Operational GENAD Course [OGC], Department of Defense, 10 February 2021), slide 3.
[70] It should be noted that sexual exploitation is also included in the definition of gender-based violence and should not be excluded from the discussion as a source of power since it is often used to manipulate or coerce victims into compliance in return for sexual favors. Lape, "Gender-Based Violence," slide 16.
[71] Massey, "Concept Plan," slide 6.
[72] Dichter, True, and Tinney, "A Call to End Gender-Based Violence in the Military."

based violence follow victims well after the incident. Victims often suffer long-term effects associated with post-traumatic stress disorder, increased risks of suicide, and other mental health and social disorders.[73]

Sexual harassment is a gateway behavior. This means that often, it has been demonstrated before the behavior further escalates to sexual assault or gender-based violence. There is a correlation between sexual harassment and an organization's environment; environments rich in dignity and respect mitigate escalation from sexual harassment to sexual assault and gender-based violence. The FHIRC report indicated that Fort Hood had a culture in which discrimination based on race and gender went unaddressed, cultivating an environment that was permissive in these behaviors and implying a culture in which sexual harassment was acceptable. In the months leading to her eventual death, Guillén indicated to family and friends that she had been sexually harassed within her unit.[74] Unfortunately, without a formal complaint, the situation escalated and ended in Guillén's death, the fatal result of gender-based violence.

Facilitating Cultural Change

Acknowledging the need for change and establishing and communicating a sense of urgency are essential to facilitating transformation within the DOD. Success requires a fundamental shift in how political leaders and military personnel approach the challenges associated with existing habits, attitudes, and beliefs. In other words, it is imperative to focus on behaviors that are unacceptable and contrary to the core American values described in the 2017 NSS and affirmed in President Biden's *Interim National Security Strategic Guidance*, that is, the universal values that "have underpinned the U.N. system" and democratic values to include equal opportunity and respect for the rule of law.[75] Military members cannot implement those values abroad if they cannot champion those values within the force's formations and communities.[76] In addition, two priorities detailed under the NSS's fourth pillar—"Advance American Influence"—are the equities of supporting individuals' dignity and empowering women and youth.[77] Fostering a culture or environment rich in dignity and respect requires that the U.S. military practice these values at home to support the first defense objective of a diverse and inclusive fighting force.[78] This also requires attention to organizational culture.

The Army People Strategy, as an example, outlines and describes culture as "the foundational values, beliefs, and behaviors that drive an organization's social envi-

---

[73] Dichter, True, and Tinney, "A Call to End Gender-Based Violence in the Military."
[74] Steinhauer, "A #MeToo Moment Emerges for Military Women after Soldier's Killing."
[75] *National Security Strategy of the United States of America*, 41; and *Interim National Security Strategic Guidance* (Washington, DC: Office of the President, White House, 2021), 13.
[76] *National Security Strategy of the United States of America*, 41.
[77] *National Security Strategy of the United States of America*, 42.
[78] *Women, Peace, and Security Strategic Framework and Implementation Plan*, 7.

ronment and [culture] plays a vital role in mission accomplishment."[79] Culture is linked to performance and enriches the foundation and values upon which military service is evaluated. However, this culture is not always positive. The converse is those aspects of culture that imply an environment that is permissive of such behaviors as racism, sexism, extremism, and incidents of sexual harassment and assault, to name a few.[80]

The sweeping change requires leader "commitment, trust, engagement and accountability."[81] Leaders can facilitate this change but need to embrace the fundamental principles that define leadership roles. For example, *Army Command Policy*, Army Regulation 600-20, specifically addresses the importance of leaders at all levels to promote a positive and constructive climate that treats soldiers and civilians with dignity and respect, provides training and professional development opportunities, and emphasizes integrity and a sense of duty.[82] *Army Leadership and the Profession*, Army Doctrine Publication (ADP) 6-22, is also relevant. It defines leadership as "the activity of influencing people by providing purpose, direction, and motivation to accomplish the mission and improve the organization."[83] As these documents underscore, leadership at all levels is critical if cultural change is to occur.

Implementing organizational change of such magnitude will have to happen from the top down and from the bottom up simultaneously. In the U.S. Army, as an example, *The Operations Process*, ADP 5-0, can direct leaders in operationalizing the strategic guidance provided in *The Army People Strategy* down to the lowest level to ensure that the messaging and vision remains unadulterated yet still adaptable to meet the challenges specific to the smallest formations.[84] Yet, DOD is well aware that the problems examined previously permeate the entire defense enterprise.

Recognizing that the U.S. military has problems with diversity and inclusion, gender-based violence, and sexual harassment and assault is the first step. However, recognition is insufficient, as the FHIRC report makes utterly apparent. The DOD must address, mitigate, and eradicate environments that seek to prevent diversity and inclusion and that foster gender-based violence, including sexual harassment and assault. Military doctrine, such as ADP 5-0, provides a foundation to assist civilian and military leaders in driving the change through an approach defined by the following six steps: 1) Understand: what is the problem? 2) Visualize: how do we fix the problem? 3) Describe: do we have a shared understanding of the problem? 4) Direct: what is the intent, and are the resources available? 5) Lead: are the leaders

---

[79] Joseph E. Escandon, "Operationalizing Culture: Addressing the Army People Crisis," *Military Review Online* (January 2021): 4.
[80] Escandon, "Operationalizing Culture," 4.
[81] *The Army People Strategy* (Washington, DC: U.S. Department of the Army, 2019), 8.
[82] *Army Command Policy*, Army Regulation 600-20 (Washington, DC: U.S. Department of the Army, 2020).
[83] *Army Leadership and the Profession*, Army Doctrine Publication 6-22 (Washington, DC: U.S. Department of the Army, 2019), 1–13.
[84] Escandon, "Operationalizing Culture," 4.

setting the standard, visible, motivated, accountable, and committed? and 6) Assess: are things improving, or do we need to do things differently?[85] Using such a structure helps leaders focus on the critical steps in promoting a change in organizational culture. Yet these steps are only one element.

A second element of facilitating a change in culture is that leaders at all levels must be able to appreciate that their understanding of organizational "culture, ethics and values" may not be interpreted the same way by those under their orders.[86] There will be other factors to consider, such as generational differences or organizational subcultures. Nonetheless, approaching change from the top and bottom simultaneously will ensure that the necessary cultural shift will reach all levels and be understood at all echelons of the organization.

Training and Leader Development

No matter the circumstance or crisis, the first response to eliminating adverse behaviors and addressing the current situation within the DOD is mandatory training. Training will support another of the department's WPS equities: inclusive leadership development. Unfortunately, to date, the emphasis of this training is more focused on getting the training done rather than the quality or effectiveness of the training. Training is rarely tailored to the demographics of the unit and is executed as a one-size-fits-all scenario.[87] When developing training packages, DOD fails to acknowledge that behavioral risks are not the same across servicemember populations.[88] Furthermore, training required on an annual basis is rarely updated and, more often than not, is unchanged from the previous year, becoming a check-the-box requirement. It is seldom presented by a subject matter expert, as well. As a result, most servicemembers have offered that the training lacks variety, fails to capture their attention, and over time desensitizes them to the topic.[89] Thus, creative and tailored training is required as current efforts favor standardization, but quality training can be ineffective if cultural change is not ongoing. The importance of training cannot be underemphasized. The FHIRC report underscored that sexual harassment and assault programs are ineffective when there is a lack of command attention to the issue or when a command climate tolerates misconduct or fails to implement protocols to minimize risk. In sum, leadership matters.[90]

Conclusion

In the end, senior leaders must take the lead in promoting a gender perspective,

---

[85] *The Operations Process*, Army Doctrine Publication 5-0 (Washington, DC: U.S. Department of the Army, 2019), 2–4.
[86] Escandon, "Operationalizing Culture," 15.
[87] Castro et al., "Sexual Assault in the Military," 6.
[88] Castro et al., "Sexual Assault in the Military," 6.
[89] Castro et al., "Sexual Assault in the Military," 2.
[90] *Report of Fort Hood Independent Review Committee*, 13, 17.

which encompasses the equities discussed, within the Services, leading by example and reinforcing gender perspectives in their organizations. However, leadership is only one component. The introduction and reinforcement of gender perspective concepts must occur at all levels of professional military education (PME) across the five armed forces. A gender perspective can be introduced as early as possible, starting with basic training for enlisted members and reinforced with the appropriate content through to education for senior noncommissioned officers. Officer education should begin in precommissioning courses and continue through senior service college.

Moreover, the training provided during PME must be tailored to the audience. If the military Services can combine leadership attention and training and act from the top and bottom of the organizational chain of command, then the likelihood of successfully implementing DOD's WPS objectives improves exponentially.

Chapter 3
A Strategic Imperative
by Commander Kristen Vechinski, USN*

> Would you make the effort to study a new perspective if you knew that you could achieve greater situational awareness than you ever had before? Or would you be outraged if an enemy discovered your neglect and used this perspective against you?
> ~Lena P. Kvarving and Rachel Grimes[91]

Women account for half the world's population; research shows that when half the population does not participate in conflict prevention and resolution, stability and prosperity are unlikely.[92] Moreover, studies reveal that including women in peace building and conflict resolution results in peace agreements that are 64 percent more likely to succeed and 35 percent more likely to last at least 15 years.[93] In 2000, the UN championed the strategic imperative of including a gender perspective in peace and security activities with the passage of UNSCR 1325, which laid the foundation for the WPS agenda. The WPS agenda defined a shift in recognizing women as not just victims of conflict but also as valuable resources and contributors to security through participation.

The WPS agenda is not about making conflict safer for women, it is about preventing and ending conflict.[94] The United States faces national security challenges with a rise in nationalism, violent extremism, and transnational criminal activity which promote violence and instability. Today's dynamic security environment demands the Joint Force be prepared to operate at all levels of competition and must not cede advantage to adversaries by failing to incorporate a gender perspective. This chapter examines how incorporating a gender perspective as a capability can improve military effectiveness by building a more lethal force and capacity which in turn can reduce operational risk.

This chapter focuses on the opportunities to maximize advantage in the security environment through the advancement of the DOD's defense objectives to promote WPS principles, providing examples of how incorporating a gender perspective im-

---

* The views expressed in this chapter are solely those of the author. They do not necessarily reflect the opinion of Marine Corps University, the U.S. Marine Corps, the U.S. Navy, the U.S. Army, U.S. Army War College, the U.S. Air Force, or the U.S. government.
[91] Lena P. Kvarving and Rachel Grimes, "Why and How Gender Is Vital to Military Operations," in *Teaching Gender in the Military: A Handbook* (Geneva, Switzerland: Geneva Centre for the Democratic Control of Armed Forces and Partnership for Peace Consortium, 2016), 1.
[92] "Preventing Conflict: The Origins of the Women, Peace and Security Agenda," in *Preventing Conflict, Transforming Justice, Securing the Peace: A Global Study on the Implementation of United Nations Security Council Resolution 1325* (New York: UN Women, 2015), 190–219.
[93] TSgt Chuck Broadway, USAF, "DoD Works to Incorporate More Gender Perspective in Operations," U.S. Department of Defense, 8 March 2018.
[94] "Preventing Conflict," 191.

proved military effectiveness in counterinsurgency operations in Afghanistan and posing challenges and recommendations based on lessons learned on how incorporating a gender perspective as an integral element of military operations can increase the ability of the Joint Force to deliver effects across the competition continuum.

Evolution

The UN developed the WPS agenda as a political framework that focuses on gender in international security based on four pillars for policymaking: prevention, protection, participation, and relief and recovery.[95] In 2000, the UN unanimously passed the landmark UNSCR 1325, the first time the Security Council addressed the disproportionate and unique impact of armed conflict on women; recognized the undervalued and underutilized contributions women make to conflict prevention, peacekeeping, conflict resolution, and peace building.[96] The UN further defined WPS concepts with seven subsequent supplemental resolutions reinforcing the importance of the WPS principles in relation to global norms and designed to influence the work of organizations supporting peace and security activities.[97]

As a result of the landmark WPS agenda, the United States published its first WPS national action plan in 2011 under Executive Order 13,595.[98] The U.S. national action plan was the first legal framework to recognize WPS as a key element of conflict prevention and resolution efforts. In 2016, a second U.S. national action plan was created, followed by the U.S. Women, Peace, and Security Act of 2017, signed into law in October 2017. Two years later, the United States published the *United States Strategy on Women, Peace, and Security*, which outlined a comprehensive whole-of-government approach to WPS.[99] The Women, Peace, and Security Act of 2017 designated DOD as a key federal organization for implementing WPS. In turn, last year DOD published the WPS SFIP, which highlights three long-term defense objectives to support the four national WPS lines of effort across 16 DOD equities supporting 7 global WPS principles.[100]

WPS—What It Is

The WPS agenda is a wide-ranging framework that outlines the role of women in the prevention and resolution of issues such as conflicts, peace negotiations, peace

---

[95] "In Focus: Women, Peace, Power," UN Women, accessed 18 January 2021.
[96] "WILPF's Women, Peace and Security Programme," Peace Women, Women's International League for Peace and Freedom, accessed 20 January 2021; and UN Security Council, Resolution 1325, Women, Peace and Security, S/RES/1325 (31 October 2000).
[97] "Guiding Documents," UN Women, accessed 18 January 2021.
[98] *United States National Action Plan on Women, Peace, and Security* (Washington, DC: Office of the President, White House, 2011); and "What Is UNSCR 1325?: An Explanation of the Landmark Resolution on Women, Peace and Security," United States Institute for Peace, accessed 20 January 2021.
[99] *United States Strategy on Women, Peace, and Security* (Washington, DC: Office of the President, White House, 2019), 4.
[100] *Women, Peace, and Security Strategic Framework and Implementation Plan* (Washington, DC: Department of Defense, 2020), 9.

building, peacekeeping, humanitarian responses, and post-conflict reconstruction. In addition, the WPS agenda emphasizes the increased participation of women and the incorporation of gender perspectives in peace and security activities.[101] In relation to the security environment, gender can be described as the term to define the social construct related to the role one learns and performs in society as both men and women.[102] Often gender issues are viewed as matters only related to women. *Gender, women,* or *sex* are not interchangeable terms.[103] Further, WPS is not a *woman's issue,* for and by women. To have a gender perspective means to observe, analyze, and understand all the roles individuals play in a culture or society.[104]

As a social construct, gender varies across cultures. From a gender perspective, war and conflict affect men, women, boys, and girls differently, which is relevant to the planning and execution of military operations. For example, adversaries can leverage gender in the security environment by weaponizing sexual violence or other forms of repression.[105] The gender ecosystem also includes men's and boys' perspectives on human security. A gender perspective takes account of the economic, political, and sociocultural constraints and opportunities of women and men.[106] To incorporate a gender perspective, the military must not simply embrace this in an internal context of inclusion and diversity but employ it externally as a capability that can have strategic, operational, and tactical impact on military operations.

WPS—What It Is Not

Critics argue WPS is a Western agenda and undermines social norms in non-Western countries by incorporating a gender perspective in military operations, which creates unintended consequences by disrupting the existing social order in the security environment. Inadequate cultural training and awareness can create pitfalls for female engagement teams.[107] Nonetheless, women from non-Western states such as Afghanistan, Sri Lanka, Syria, and Libya, risk their lives to resolve conflict and promote peace.[108] To suggest women involved in peace and security is a Western agenda undermines their contributions to promote stability. For example, the female Kurdish People Protection Units in Syria were instrumental in increasing military effectiveness during the war with the Islamic State of Iraq and Syria (ISIS) near the northern Syrian border. The female-led units executed operations to reclaim key terrain such as

---

[101] Joan Johnson-Freese, *Women, Peace and Security: An Introduction* (New York: Routledge, 2019), 9.
[102] Kvarving and Grimes, "Why and How Gender Is Vital to Military Operations," 2.
[103] Joana Cook, *A Woman's Place: US Counterterrorism since 9/11* (New York: Oxford University Press, 2020), 5.
[104] Col Veronica Oswald-Hurtkay, USA, email message to author, 9 February 2021.
[105] Laura Sjoberg, *Gender, War, and Conflict* (Cambridge, United Kingdom: Polity Press, 2014), 41–42.
[106] Cook, *A Woman's Place,* 5.
[107] Richard Ledet, Pete A. Turner, and Sharon Emeigh, "Recognizing the Ethical Pitfalls of Female Engagement in Conflict Zones," *Journal of Military Ethics* 17, no. 4 (March 2018): 201, https://doi.org/10.1080/15027570.2019.1585619.
[108] Sjoberg, *Gender, War, and Conflict,* 40.

Raqqa, the former capital of the Islamic State in Syria.[109] The female militia members served as a force multiplier that increased lethality and built capacity in their military operations against ISIS. Their participation in security operations also served to promote social and cultural advancement of Kurdish women in their society.

WPS skeptics question why women matter in security operations when it is primarily men who engage in conflict. While it is men who overwhelmingly participate in war and conflict, the number of women soldiers, insurgents, and terrorists is on the rise, in part because of the specific advantages in weaponizing women.[110] Terrorist groups or insurgent groups may use women in suicide bombing missions because they are less likely to be scrutinized than men. For instance, in the Vietnam Conflict, the Viet Cong sent women into the jungle holding cluster bombs disguised as babies.[111] In violent extremist organizations, women's role as terrorist actors in support of political violence has increased. Whether in a traditional combat role or being leveraged for aspects of their gender role in a particular environment, women have always been actors in conflict.[112] Not all women are peaceful, but overwhelmingly women are the first to mobilize for peace and reconciliation.

Therefore, incorporating WPS principles can be viewed as a complement to other policies and strategies of security partners and supporting institutions such as the UN and NATO. Additionally, others believe that WPS equates to promoting the concept of gender neutrality or more specifically gender blindness. *Gender blindness* is defined as the lack of awareness of how men and women are differently affected by a situation because of their respective roles, status, and priorities in their societies.[113] However, gender blindness in military operations is counter to improved effectiveness as it discounts the various interests, threats, or needs of women, men, boys, and girls in a security environment. International organizations such as the World Bank have collected empirical evidence that suggests a gender-blind approach to peace and security will lead to failure and increase instability and violence.[114]

WPS—Why It Matters

The 2017 NSS emphasizes terrorists and transnational criminal organizations "prey on the vulnerable" and actively compete against the United States, its allies, and

---

[109] Gayle Tzemach Lemmon, *The Daughters of Kobani: A Story of Rebellion, Courage, and Justice* (New York: Penguin, 2021), 160.
[110] Jamille Bigio and Rebecca Turkington, "U.S. Counterterrorism's Big Blindspot: Women," *New Republic*, 27 March 2019.
[111] Sjoberg, *Gender, War, and Conflict*, 42.
[112] Cook, *A Woman's Place*, 2–4.
[113] Sahana Dharmapuri, "Back to the Basics: Gender Blindness Negatively Impacts Security," *Our Secure Future* (blog), 3 March 2017.
[114] Robert Egnell and Mayesha Alam, "Introduction: Gender and Women in the Military—Setting the Stage," in *Women and Gender Perspectives in the Military: An International Comparison*, ed. Egnell and Alam (Washington, DC: Georgetown University Press, 2019), 13–14.

its partners.[115] Malign actors pose a threat to national security and human security. Implementing WPS principles in the operational environment is generally considered a soft power approach to security challenges. Soft power tools are persuasive and population-focused that can help shape, influence, or stabilize the environment.[116] Furthermore, the definition of security evolved to emphasize human security where the social and human terrain is as important as the land terrain. Frank Hoffman and Michael Davies emphasize this point with their assertion that "to succeed, Joint commanders must be able to successfully maneuver in the most decisive domain, and that is the Human Doman."[117] A human security approach requires the understanding of particular threats to a particular group in order for the group of people to have freedom from fear or freedom from want. Naval War College professor Joan Johnson-Freese further highlights the shift in priorities in the security landscape from conventional warfare to irregular warfare where the population is essential to achieving political aims.[118] Similarly, the UN shifted from the state-centric focus of safety from military aggression to one that is centered on the security of individuals and their protection and empowerment.[119] The core of operationalizing WPS principles as a capability to build capacity and increase lethality depends on understanding the forces and trends in the security environment from different perspectives.

The passing of UNSCR 1325 through significant research established the strategic value of a gender perspective in peace and security affairs.[120] The military plays a key role as the organization to operationalize WPS principles in security affairs. In turn, WPS enables the military to maximize its effectiveness in support of national policy goals. In the most basic terms, an effective military succeeds by performing the tasks its political leadership asks of it.[121] According to the *Joint Operations*, Joint Publication (JP) 3-0, effective operations require an understanding of the multidimensional battlespace, including the "interrelationship of the informational, physical, and human aspects that are shared by the OE [operational environment] and information environment."[122] There is evidence that incorporating a gender perspective in operations in Afghanistan contributed to improving the effectiveness of ground combat and Special Operations Forces (SOF) missions.

---

[115] *National Security Strategy of the United States of America* (Washington, DC: Office of the President, White House, 2017), 10.
[116] Cook, *A Woman's Place*, 3.
[117] Frank Hoffman and Michael C. Davies, "Joint Force 2020 and the Human Domain: Time for a New Conceptual Framework?," *Small Wars Journal*, 10 June 2013, as quoted in Col Joseph D. Celeski (USA, Ret), "SOF, the Human Domain and the Conduct of Campaigns," *Special Warfare* 27, no. 3 (July–September 2014): 5.
[118] Johnson-Freese, *Women, Peace and Security*, 14.
[119] Sabrina Karim, "Women in UN Peacekeeping Operations," in *Women and Gender Perspectives in the Military*, 24.
[120] Robert Egnell, "Gender Perspectives and Military Effectiveness: Implementing UNSCR 1325 and the National Action Plan on Women, Peace, and Security," Inclusive Security, March 2016.
[121] Egnell, "Gender Perspectives and Military Effectiveness."
[122] *Joint Operations*, Joint Publication 3-0 (Washington, DC: Joint Chiefs of Staff, 2018), iv-1, iv-2.

WPS Done Right—Afghanistan

The long wars in Afghanistan and Iraq highlight the population's significance in the shift from conventional to irregular warfare.[123] Countering violent extremism and insurgency requires more than a traditional military response. General Rupert Smith's theory, "war amongst people," states that modern conflict is unlikely to be resolved by force alone, in particular when complex socioeconomic and political problems exist in the security environment.[124] Incorporating a gender perspective as a strategic imperative underscores the relevance of a population centric focus on warfare whether categorized as irregular, hybrid, or new generation warfare. In Afghanistan, the incorporation of a gender perspective as a capability in military operations enhanced operational effectiveness and reduced risk with enhanced situational awareness.

To illustrate, during the insurgency in Iraq, U.S. Marine Corps units created the female Lioness Teams as special cordon and search teams to fill a security gap and improve military effectiveness by mitigating the security limitations of searching the local female population.[125] The culturally sensitive search methods in a gender-segregated area of operations allowed military forces to identify females who were active participants in the conflict and in some cases male insurgents who used full traditional female garments to avoid detection.[126] The complex and asymmetrical threats of the battlespace required an expanded security capacity of forces to have access to all aspects of the population. The Lioness Teams joined male Marines and Army soldiers on raids, security patrols, and security checkpoints intended to search for weapons and explosive vests.[127] Using female troops in a population of neutral citizens, including women, children, and the elderly, can create conditions to leverage the population for support. By leveraging both women and men, a counterinsurgency operation has the potential to disrupt insurgent activity and deny a base of support to the adversary.[128]

Similar to Iraq, given Afghanistan's gender-specific cultural and social norms, military units incorporated a gender perspective as a capability in order to increase military effectiveness with the growing counterinsurgency operations. In 2009, Female Engagement Teams (FETs) in Afghanistan evolved from the success of the Lioness Teams in Iraq as an operational innovation to increase security and information operations and to build credibility with the local population.[129] In these gender-segregated

---

[123] Johnson-Freese, *Women, Peace and Security*, 15.
[124] Marcelo O. L. Serrano, "War Amongst the People, or Just Irregular," *Small Wars Journal*, 25 March 2014.
[125] Cook, *A Woman's Place*, 30.
[126] Megan Katt, "Blurred Lines: Cultural Support Teams in Afghanistan," *Joint Forces Quarterly* 75, no. 4 (2014): 107.
[127] Gayle Tzemach Lemmon, *Ashley's War: The Untold Story of a Team of Women Soldiers on the Special Ops Battlefield* (New York: Harper Perennial, 2016), 8–10.
[128] Frank Gasket, Ryan Voneida, and Ken Goedecke, "Unique Capabilities of Women in Special Operations Forces," *Special Operations Journal* 1, no. 2 (November 2015): 105–11.
[129] WO1 Raymond T. Kareko, "Female Engagement Teams," *NCO Journal* (October 2019): 1–5.

communities, the teams created access to the female population without increasing tension with local Afghan males, in particular village elders who held considerable sway over village politics.[130] Information provided by local village women proved useful; more importantly, the village elders willingly gave their approval since the FETs did not violate their cultural rules.[131] A successful counterinsurgency operation must have a comprehensive knowledge of the society and culture. The inclusion of a gender perspective as capability not only improved information gathering about the communities where U.S. and NATO troops were operating, it enhanced the credibility of the military operations with the local population.[132]

The FETs served a particular operational role in the counterinsurgency campaign. As such, the FETs embodied the definition of *enabler*: "an organization or capability that supports a particular course of action and/or accomplishment of a particular objective."[133] In direct support of companies or battalions, the female enablers engaged Afghan men and women to leverage the population in accordance with operational objectives in all phases of counterinsurgency operations.[134] FET employment built capacity in information dissemination, medical outreach and education, security support, and civil-military activities.[135] Further, access to the population allowed the FETs to gather useful intelligence about Taliban and al-Qaeda positioning based on input from local Afghan women. The passive information collection built capacity through improved human terrain understanding.[136] Moreover, the FETs provided an opportunity to leverage a gender focus for cooperation with the local population, operational partners, and international organizations in the complex multidimensional battlespace of Afghanistan.[137] Female enablers in military operations can foster cooperation and respect with the local population across families and social networks, which can be a powerful tool when operating in the human terrain to counter violent extremism.

Furthermore, Afghan men often viewed female soldiers as a third gender, which allowed them to interact with all aspects of the population since the elders did not hold them to the same standards as local women.[138] Confronting threats in the contempo-

---

[130] Cook, A Woman's Place, 33–34.
[131] Ledet, Turner, and Emeigh, "Recognizing the Ethical Pitfalls of Female Engagement in Conflict Zones," 204–5.
[132] Brigitte Rohwerder, Lessons from Female Engagement Teams: GSDRC Helpdesk Research Report 1186 (Birmingham, United Kingdom: Governance and Social Development Resource Centre, University of Birmingham, 2015), 2–7.
[133] Capt Colin Marcum, "How Enablers Shape the Deep Fight for the BCT," reprint from Fires (March–April 2017): 3.
[134] 1stLt Zoe Bedell, "United States Marine Corps Female Engagement Team" (PowerPoint presentation, NATO, May 2011), slide 6.
[135] Bedell, "United States Marine Corps Female Engagement Teams."
[136] Sjoberg, Gender, War, and Conflict, 42.
[137] Brenda Oppermann, "Women and Gender in the US Military: A Slow Process of Integration," in Women and Gender Perspectives in the Military, 120–21.
[138] Jessica Glicken Turnley, "Funhouse Mirrors: Reflections of Females in Special Operations Forces," Special Operations Journal 5, no. 1 (2019): 28–29.

rary security environment requires intervention with both combatants and noncombatants to achieve effects. Incorporating a gender perspective through FET employment increased lethality in security missions and built capacity in women's governance (via *shuras*, or councils, and medical outreach events).[139] In a counterinsurgency environment with the population as the center of gravity, mission success depends on the ability to engage and influence the population within complex cultural norms and societal roles.[140] By 2011, NATO's International Security Assistance Force "stated that the [FETs] were 'battlefield enablers that influence [and] inform'."[141]

Founded in the accomplishments of Lioness Teams and FETs, U.S. Special Operations Command (SOCOM) established Cultural Support Teams (CSTs) to support similar humanitarian activities and required search skills.[142] The CST model was not as structured as the FET model. The key difference between the two is that FETs were used for activities to "soften the footprint" of coalition forces while the intent of CSTs was to build capacity for persistent presence and engagement in vulnerable Afghan villages.[143] The CST concept included support of counterterrorism raids and village stability operations (VSO) conducted with SOF personnel as part of the counterinsurgency campaign in Afghanistan.[144] SOCOM commander Admiral Eric Olson championed the all-female team concept. Admiral Olson asserted that the United States needed to be agile, adaptive, and innovative to be effective.[145] Olson built on the idea to incorporate a gender perspective in the special operations missions to "increase the team's ability to assess cultural climate and understand the local environment."[146]

To increase the lethality of their teams and offer affective options for commanders, Olson argued with senior leaders that female enablers created a capability worth building. Olson acknowledged that "America wasn't and isn't going to kill its way to the end of its post–9/11 wars."[147] While Olson met with organizational resistance within DOD, Joint Special Operations Command (JSOC) commander Admiral William McRaven supported the assertion that female enablers could make U.S. Army Ranger missions more successful.[148] Both leaders agreed the counterinsurgency mission would not get done if the SOF teams did not have access to half of the Afghanistan population.

---

[139] LtCol Janet R. Holliday, USA, "Female Engagement Teams: The Need to Standardize Training," *Military Review* 92, no 2 (March–April 2012): 93.
[140] Cook, *A Woman's Place*, 33–34.
[141] Eileen Rivers, *Beyond the Call: Three Women on the Front Lines in Afghanistan* (New York: Da Capo Press, 2018), 33; and Maj Sheila Medeiros, *ISAF Joint Command's Female Engagement Team Program: Comprehensive Assessment Report* (Kabul, Afghanistan: ISAF Joint Command, 2012).
[142] Oppermann, "Women and Gender in the US Military," 120; and Adm Eric T. Olson, USN (Ret), "On the Original Role and Scope of Females in Cultural Support Teams," SOFX.com, 3 December 2016.
[143] Katt, "Blurred Lines," 109.
[144] Lemmon, *Ashley's War*, 13–14.
[145] Lemmon, *Ashley's War*, 12–13.
[146] Cook, *A Woman's Place*, 187.
[147] Gayle Tzemach Lemmon, "The Army's All-Women Special Operations Teams Show Us How We Will Win Tomorrow's Wars," *Washington Post*, 19 May 2015.
[148] Lemmon, *Ashley's War*, 14–15.

JSOC's request for forces for female soldiers to serve alongside the 75th Ranger Regiment during night raids came from the idea of a strategy to make their mission more effective, not by means of gender inclusion as equality.[149] Correspondingly, senior leaders made the choice of avoiding the "female" label in the branding of the capability and used cultural support to identify the opportunity for the CSTs to access the people and places where all-male units could not. SOCOM leadership recognized that regardless of how proficient SOF teams were at targeting insurgents, teams needed to adopt a gender perspective to understand and develop trust with the local population.[150]

CSTs demonstrated SOCOM's strategic approach to improve operational effectiveness of the night raids and VSO missions by incorporating a gender perspective in the security environment. The traditional SOF missions, such as counterinsurgency, stability operations, and foreign internal defense, form the foundation of the VSO mission.[151] The CSTs increased the lethality through their ability to facilitate communication with the local population when conducting sensitive operations.[152] Incorporating a gender perspective can expand the battlespace to include all of the population through WPS engagement activities in security, governance, and development.[153] Army SOCOM commander Lieutenant General John Mulholland reinforced the capability CSTs brought to the fight as a force multiplier through incorporating a gender perspective in the complex security environment. Referencing the CST impact on operations in Afghanistan, he remarked, "Make no mistake about it, these women are warriors; these are great women who have also provided enormous operational success to us on the battlefield by virtue of their being able to contact half the population that we normally do not interact with."[154]

Colonel Mark O'Donnell of the 75th Ranger Regiment also supported how CSTs increased effectiveness of information gathering, asserting, "From an intelligence standpoint what they provide by engaging women and children on the objective contributes immeasurably to our success."[155] The ability to gather information about terrorist activity from talking to all of the local people allowed the female enablers to pick up crucial information on patterns of behavior. In building trust, the female enablers had access to immediate and actionable information.[156]

Research from Carnegie Mellon University and Massachusetts Institute of Technology indicates that a group's collective intelligence may increase as the percentage

---

[149] Lemmon, *Ashley's War*, 15.
[150] Lemmon, "The Army's All-Women Special Operations Teams Show Us How We Will Win Tomorrow's Wars"; and Jared M. Tracy, "The U.S. Army's Cultural Support Team Program: Historical Timeline," *Veritas* 12, no. 2 (2016).
[151] Katt, "Blurred Lines," 107.
[152] Cook, *A Woman's Place*, 187.
[153] Cook, *A Woman's Place*, 188.
[154] Lemmon, *Ashley's War*, 257.
[155] Lemmon, *Ashley's War*, 259.
[156] Rivers, *Beyond the Call*, xxi.

of females in the group increases due to women's "social sensitivity" in reading other people's emotions, a key skill when gathering information.[157] In sum, lessons learned from the war in Afghanistan demonstrated military forces cannot always execute all the critical supporting functions without a gender perspective incorporated into military operations. More importantly, incorporating gender perspective as a capability improved operational effectiveness through improved passive intelligence gathering, enhanced legitimacy, and better force protection by accessing all of the Afghan population.

As demonstrated through the operational innovations in Afghanistan, the successful implementation of WPS principles hinges on the flexibility and adaptability of military operations. To date, the DOD has engaged with more than 50 partner nations to demonstrate the significance of women's meaningful participation in national security as well as share best practices on the recruitment, retention, and employment of women in the military forces to increase interoperability with many U.S. security partners.[158] WPS needs are different in the various areas of operations as highlighted by the diverse WPS activities of geographic combatant commands.

According to Cori Fleser, a WPS advisor on the Joint Staff, combatant commands and their components have advanced DOD's implementation by incorporating gender and human security issues in campaign plans, security cooperation, and exercises and training. For example, U.S. Indo-Pacific Command's (INDOPACOM) WPS office conducted workshops and disaster relief exercises led by U.S. Army Pacific to improve WPS integration in humanitarian and disaster response support. U.S. Southern Command's WPS program supports women's participation in security matters during its key leader engagements with strategic partners in their areas of responsibility, while U.S. European Command directs its efforts to an interagency approach to WPS implementation by supporting NATO. Finally, U.S. Africa Command successfully incorporated WPS principles in its peacekeeping activities to build capacity with African troop-contributing countries.[159]

Utilizing a gender perspective to their advantage, terrorist organizations like ISIS often attempt to exploit gender norms to gain support from local populations and increase instability. Local women's organizations are repeatedly stakeholders in increasing resilience within the community to counter violent extremism. Recognizing the link between gender perspective and military effectiveness, U.S. Special Operations Command Africa highlighted the role of female security forces and violent extremist organizations' exploitation of gender dynamics during their 2017–19 Flint-

---

[157] Frank Gaska, Ryan Voneida, and Ken Goedecke, "Unique Capabilities of Women in Special Operations Forces," *Special Operations Journal* 1, no. 2 (November 2015): 109, https://doi.org/10.1080/23296151.2015.1070613; and Anita Williams Woolley et al., "Evidence for a Collective Intelligence Factor in the Performance of Human Groups," *Science* 330 (October 2010): 686–88.
[158] Jim Garamone, "Women, Security, Peace Initiative Militarily Effective," U.S. Department of Defense, 5 November 2020.
[159] Garamone, "Women, Security, Peace Initiative Militarily Effective."

lock Exercises.[160] The INDOPACOM gender advisor acknowledges that WPS has evolved from a smart power asset to a learned and applied capability.[161]

Adding a gender perspective to military operations not only has the power to elevate the strategic appraisal of how to best use the military instrument to support national security objectives to achieve political goals, it can improve the military effectiveness to achieve those goals. Countering violent extremist organizations (CVEO) has the opportunity to use a gender perspective in the security environment as a capability against armed nonstate actors.[162] CVEO will likely remain an enduring theme in the global security landscape with an increase in right-wing extremism and jihadism. Malign actors such as ISIS affiliates not only spoil peace building and stability, they complicate the human security environment as a strategic challenge to vulnerable states and the allies working with them. As noted, leaders can tailor WPS on the tactical level or incorporate it into regional efforts or theater perspective.

Way Ahead

The operational capabilities demonstrated by FETs and CSTs in Afghanistan should not be viewed as just a U.S. Central Command area of responsibility requirement because of the local culture's strict gender separation. Military operations can employ a gender perspective as a capability in any security environment. When the strategy involves engaging the local population, a gender analysis of the environment is essential.[163] Implementing WPS principles in future contingencies, to include nonlethal military operations, will likely be necessary to engage with all aspects of the local population. Gender perspective as a capability can allow units to gain better acceptance from the local population and increase tangible information collection to enhance operations.[164]

To maximize the effectiveness of incorporating a gender perspective in the operational environment, implementation of WPS principles should be proactive and not reactive. Effectively assessing and operating in the human terrain is necessary for success in a spectrum of activities: stability, peace, humanitarian, and CVEO operations.[165] The military serves as an integral supporting national instrument in achieving national policy goals to deter aggression, disrupt al-Qaeda and related terrorist networks, and prevent an ISIS resurgence.[166] Integrating WPS principles into conflict prevention and stability operations not only advances important national interests to break the cycle of fragility and promote peaceful self-reliant nations,

---

[160] Garamone, "Women, Security, Peace Initiative Militarily Effective."
[161] Sharon Goveia Feist, email message to author, 19 April 2021.
[162] Cook, *A Woman's Place*, 418–20.
[163] Kvarving and Grimes, "Why and How Gender Is Vital to Military Operations," 14.
[164] Cook, *A Woman's Place*, 204.
[165] Cook, *A Woman's Place*, 421.
[166] *Interim National Security Strategic Guidance* (Washington, DC: Office of the President, White House, 2021), 11.

incorporating a gender perspective can improve effectiveness in counterinsurgency operations. Therefore, WPS should not be categorized as a unique consideration, as outlined in *Stability Operations*, Army Doctrine Publication (ADP) 3-07.[167] Instead, it should be considered a necessary element of a comprehensive approach to the training, planning, and execution of military operations. WPS is not the singular answer to the multifaceted problems that exist in the global security environment. However, incorporating a gender perspective can be a force multiplier in the security sector and it can increase the peacekeeping capacity of interagency, nongovernmental organizations, and host nation partners.

Challenges

When leaders incorporate a gender perspective into mission and define clear objectives, WPS principles can be operationalized beyond the incremental success. Empirical evidence and interviews with former FET members underscore how the often-ad hoc nature of FET projects or lack of consistency with FET engagement undermined their utility.[168] For instance, the lack of coordination during the relief in place/transfer of authority between units sidelined progress on projects and key relationship networks that took months to rebuild.[169] Efforts for sustainable implementation of WPS principles require that the military personnel tasked with executing the WPS-related activities have relevant and standardized training. Additionally, both military and civilian leadership must support and value the incorporation of a gender perspective in military operations beyond the legal mandate. Moreover, the military must move beyond incorporating a gender perspective as simply female engagement or a special program nested inside military activities.

NATO secretary general's special representative for WPS, Ambassador Marriet Schuurman, asserted in a 2017 interview that operations in Afghanistan demonstrated the relevance of gender perspective to improve strategic awareness, mitigate harmful consequences, and contribute to lasting peace.[170] NATO considers WPS principles as part of its core planning process and a capability beyond just integrating women—it is about how a gender perspective contributes to a more effective military team.[171] Critics argue there is not enough evidence to truly establish women's contribution to peace and security efforts in Afghanistan. However, the research that forms the very foundation of UNSCR 1325 and follow-on resolutions related to WPS establishes the value of gender perspective in both peace operations and military affairs.[172]

---

[167] *Stability*, Army Doctrine Publication 3-07 (Washington, DC: Department of the Army, 2019), 3-1–3-13.
[168] Rohwerder, *Lessons from Female Engagement Teams*, 3–5.
[169] Holliday, "Female Engagement Teams," 93.
[170] "The Women, Peace and Security Agenda: Integrating a Gendered Perspective into Security Operations," interview with Amb Marriët Schuurman, in *Fletcher Forum of World Affairs* 41, no. 1 (Winter 2017): 106.
[171] "Women, Peace and Security," North Atlantic Treaty Organization, last modified 14 September 2021.
[172] Egnell, "Gender Perspectives and Military Effectiveness," 74–75.

WPS Implementation

To achieve effective use of a gender perspective as a capability, WPS principles should be institutionalized and elevated from a special area of professional military education. Across all levels of Service schools, leadership needs to include the overarching WPS Agenda and subsequent DOD SFIP into applicable core curriculums. At the same time, bottom-up awareness and training can be readily introduced through existing general military training (GMT) topics linked to WPS pillars such as mandatory GMT on inclusivity and diversity, sexual assault, and trafficking-in-persons. The institutionalization of WPS principles within the military organization GMT structure assists in anchoring WPS in general knowledge, training, and application across the entire organization, not just with members tasked with working WPS implementation.

Since the signing of UNSCR 1325, DOD has spent 20 years developing its SFIP for advancing WPS principles. The fiscal year 2021 National Defense Authorization Act mandates that the secretary of defense lead the DOD effort to implement the DOD WPS SFIP no later than September 2025.[173] While the law requires an SFIP, DOD must not miss critical opportunities to implement a gender perspective in future military operations to improve effectiveness. Enabling a gender perspective allows opportunities to tailor military operations and techniques to harness the energy of the population in demanding security environments with competing interests.

WPS principles of prevention, protection, and participation are constants, but how a strategy incorporates them varies. Senior military leaders should consider WPS principles in planning and executing operations that can ultimately improve mission effectiveness. NATO embraced the WPS agenda when it assigned a gender advisor (GENAD) at the strategic, operational, and tactical levels in Afghanistan.[174] Moreover, NATO recognizes gender analysis as a capability that can increase the possibility to reveal challenges and opportunities to the mission objective.[175]

WPS principles intersect and support several activities in the range of military operations including, but not limited to, humanitarian engagements, security cooperation, peace operations, and CVEO activities.[176] When women are viewed as resources and not as victims, they can be active participants and contributors to their own security. The DOD has taken several steps to institutionalize WPS principles into guiding documents and activities. Currently, WPS principles are referenced in multiple national level strategies, Joint publications, and geographic combatant command guidance.

However, successful implementation of these principles will rely on not only institutionalizing the DOD WPS SFIP through policy and strategic guidance, but from the

---

[173] William M. (Mac) Thornberry National Defense Authorization Act for Fiscal Year 2021, Pub. L. No. 116-283 (2020).
[174] Lisa A. Aronsson, "Listen to Women: Diversity, Equity, and Inclusion," Atlantic Council, 14 October 2020.
[175] Clare Hutchinson, "NATO Statement at the United National Security Council Open Debate on Women, Peace and Security," North Atlantic Treaty Organization, last updated 3 November 2020.
[176] Cook, A Woman's Place, 32–34.

efforts of top-down and bottom-up proactive leaders who work to promote change from within as champions of WPS principles. Proactive senior military leaders understand the international and national frameworks that support the WPS agenda. They also recognize operational opportunities and institutional challenges within the military organization. Admirals Olson and McRaven pursued groundbreaking solutions to address the complex problems of the Afghanistan counterinsurgency campaign. They established the CSTs for their operational capabilities. Continued progress requires forward-thinking innovative senior leadership to implement the WPS agenda into action for measurable success within applicable DOD equities.

Further, effective WPS implementation requires male champions to achieve a cultural shift within the typically male-dominated security organizations. Support from senior male leaders can help overcome the resistance and skepticism of WPS principles' operational effectiveness.[177] Senior leaders can cultivate champions of WPS inside the military organization to anchor new approaches. Champions can then advocate for the operationalization of WPS principles as an external and internal priority. As illustrated from experiences in Afghanistan, incorporating WPS principles can be a force multiplier in the battlespace, but when leaders treat gender perspective as an ad hoc afterthought, success will remain elusive.

Conclusion

The question remains on how to close the gap between the WPS framework and the operational environment. While there has been progress in implementing the WPS agenda from the top down, the DOD needs to accelerate its efforts to incorporate a gender perspective in doctrine at the strategic, operational, and tactical level as an integral element of military training, planning, and operations. An effective U.S. security strategy must be a cooperative whole-of-government approach with partners and allies to create conditions for long-term stability, which is core to the DOD WPS Strategy line of efforts and equities. The failure of the Joint Force to use gender perspective as a capability, when and where it is applicable, poses a risk to maximizing operational effectiveness.

The Biden administration's 2021 *Interim National Security Strategic Guidance* does not call out the WPS agenda specifically, however, it calls for a "new and broader understanding of national security."[178] In an era of constrained and declining budgets, implementing WPS principles provides a low-cost investment with potential to yield measurable dividends. Increasing effectiveness by adding gender perspective as a capability to military operations offers new ways to think through or approach multidimensional security problems. As illustrated through the examples of female engagement teams in Afghanistan, enabling a gender perspective allows

---

[177] Erin Cooper, "Mobilizing Men as Partners for Change," Our Secure Future, 22 June 2020.
[178] Biden, *Interim National Security Strategic Guidance*, 6–7.

the opportunity for senior leaders to tailor responses and techniques in a volatile and complex security environment.

As a leader in the global security environment, the United States has a responsibility to lead from the front in its alliances and partnerships to build lasting security relationships, which includes putting into action the principles of UNSCR 1325. Similarly, as the United States reengages with international institutions and modernizes partnerships around the world, incorporating a gender perspective in support of DOD equities supports shared norms and human dignity. The military can capitalize on this valuable tool as a means of effectiveness as a force multiplier in operational planning and execution in three principal ways: improved information gathering, enhanced credibility, and better force protection by reaching all of the population.[179] Incorporating gender perspectives in all phases of an operation as a strategic imperative, should be both a goal and means to improving the effectiveness of military operations to support the political objectives of US national policy.

---

[179] Sahana Dharmapuri, "Just Add Women and Stir?," *Parameters* 41, no. 1 (Spring 2011): 59–61.

Chapter 4
A Kotter Approach for Geographic Combatant Commands
by Colonel Steven J. Siemonsma, ARNG*

> The prevention of conflict, the protection of human rights, and the promotion of peace and security worldwide cannot be achieved without the full and equal participation of women.
> ~Ambassador, Melanne Verveer[180]

Taking steps to address gender inequality, the UN Security Council in October 2000 passed a resolution calling on nations to mainstream gender perspectives—how persons are treated according to their gender—within global peace and security operations.[181] In response, the United States mandated that its key agencies incorporate gender perspectives. These included DOD and its geographic combatant commands (GCCs), which are arguably DOD's face to the world. For various historical and social reasons, GCCs do not yet reflect the United States's current gender composition.[182] Instead, they are heavily male, despite an enduring value of equality. To close any real or perceived say-do gap between U.S. domestic equality rhetoric and the gender equality it encourages abroad, GCCs should—and now must—model gender perspective principles. This chapter employs John Kotter's Leading Change framework to suggest approaches for incorporating a gender perspective at the GCC level and to analyze how GCCs might apply elements of Kotter's theory to incorporate gender perspective into their organizational cultures.

Background

UNSCR 1325 identified an urgent need to mainstream gender perspectives into global peace and security operations.[183] Gender perspective is broadly defined as the treatment of a man or woman in society and how that treatment sculpts a person's intrinsic needs, which include their interests, financial means, and security.[184]

In 2011, to support gender perspective in the United States, President Barack H. Obama signed Executive Order 13,595, mandating the first-ever U.S. *National Action Plan on Women, Peace, and Security*. This order noted that "promoting wom-

---

* The views expressed in this chapter are solely those of the author. They do not necessarily reflect the opinion of Marine Corps University, the U.S. Marine Corps, the U.S. Navy, the U.S. Army, U.S. Army War College, the U.S. Air Force, or the U.S. government.
[180] Melanne Verveer, foreword to *Women and Gender Perspectives in the Military: An International Comparison*, ed. Robert Egnell and Mayesha Alam (Washington, DC: Georgetown University Press, 2019), vii.
[181] UN Security Council, Resolution 1325, Women, Peace and Security, S/RES/1325 (31 October 2000).
[182] Doug K. Serota, email message to author, 26 February 2021.
[183] UNSCR 1325.
[184] JKO WPS training courses.

en's participation in conflict prevention, management, and resolution, as well as in post-conflict relief and recovery, advances peace, national security, economic and social development, and international cooperation."[185]

On 6 October 2017, President Donald J. Trump signed the Women, Peace, and Security Act of 2017 into law.[186] The act instructed DOD to develop an implementation strategy for improving women's meaningful participation in peace and security processes across the Joint Force.[187]

In 2017, Congress further mandated that key departments and agencies such as the DOD implement strategies for improving women's meaningful participation in peace and security processes.[188] In June 2019, in addition to the 2017 Act and Congress's mandate, the U.S. government published the *United States Strategy on Women, Peace, and Security*.[189] Finally, in response to the above governmental mandates, DOD then published its own WPS SFIP in June 2020.[190] DOD's plan committed it to "support the intent of the . . . [Act] through attention to the composition of its personnel and the development of its policies, plans, doctrine, training, education, operations, and exercises."[191]

Authorization and funding for changes came with the passage of H.R. 6395, the National Defense Authorization Act (NDAA) of January 2021. The NDAA mandated broadly that the secretary of defense lead the DOD to implement three objectives, with an end date of 30 September 2025.

> First, it must implement the 2020 SFIP. Second, it must establish policies and programs which support the SFIP's objectives. Third, it must ensure that sufficient personnel are trained to advance the SFIP's objectives.[192]

Broadly speaking, these mandates mean three things, as stated in the earlier government guidance cited above. First, that DOD will allow for women's "meaningful participation across . . . the Joint Force"; second, that the DOD will support the meaningful participation of women in partner nations in defense and security sectors; and third, that partner nations' defense and security sectors would protect females' safety, security, and human rights.[193]

---

[185] *Executive Order 13,595—Instituting a National Action Plan on Women, Peace, and Security* (Washington, DC: White House, 2011).
[186] Women, Peace, and Security Act of 2017, Pub. L. No. 115–68 (2017).
[187] Women, Peace, and Security Act of 2017, Pub. L. No. 115–68 (2017).
[188] Women, Peace, and Security Act of 2017, Pub. L. No. 115–68 (2017).
[189] *United States Strategy on Women, Peace, and Security* (Washington, DC: Office of the President, White House, 2019).
[190] *Women, Peace, and Security Strategic Framework and Implementation Plan* (Washington, DC: Department of Defense, 2020).
[191] *Women, Peace, and Security Strategic Framework and Implementation Plan*, 11.
[192] William M. (Mac) Thornberry National Defense Authorization Act for Fiscal Year 2021, Pub. L. No. 116–283 (2020).
[193] *Women, Peace, and Security Strategic Framework and Implementation Plan*, 7.

The first of these mandates pertains to the Joint Force and is therefore internal. It will affect the DOD's GCCs, that is, its boots-on-the-ground forces, which are overwhelmingly male-gendered (though slightly more than half the U.S. population is female).[194]

More than most forces, GCCs project, and therefore reflect, not only the U.S. military's unbalanced gender composition but also its gender perspective—through daily action abroad, before an external audience. What the GCCs' global observers see are U.S. forces led and managed by males, even as America censures some nations for gender inequality in their institutions.[195] To avoid a real or perceived "say-do" gap, U.S. GCCs should model gender perspective principles.

Assumptions

Recent U.S. government mandates and plans to leverage WPS principles imply these are not yet institutionalized across its departments and agencies. This research project assumes GCCs have therefore not fully incorporated gender perspective.[196] It further assumes that doing so will involve organizational culture changes. These may involve intrinsic needs, such as personal interests and security, but not financial security, as gender does not affect U.S. military members' salaries. Gender-related promotion potential affecting earnings within a GCC context lies outside the scope of this research. How different missions may affect different GCCs' cultures and gender mixes may be a topic for further research; however, for the purposes of this chapter, the authors assume only that some or all U.S. GCCs have not yet fully implemented gender perspective.

Other assumptions relate to problems GCCs may face as they incorporate gender perspective. First, DOD's five-year implementation timeline may be too aggressive.[197] This assumption rests on the example of GCCs' implementation of the multidomain operations concept, launched in April 2015.[198] As of 2021, the multidomain operations concept is still a work in progress, six years on. One may assume that implementing gender perspective as another high-priority form of integration may take no less time.

To incorporate a gender perspective in a given GCC, a second assumption will be that DOD implementation orders and an NDAA authorization for the SFIP implementation alone cannot lead to change. Rather, the authors assume change is possible only with leadership by executive agents at the operational echelon. However, none has thus far been assigned to oversee GCC-level gender perspective implementation. This leads to a third assumption, that a lack of specificity in the SFIP as to who should lead

---

[194] Female persons, percent, "Quick Facts, United States," U.S. Census Bureau, accessed 30 January 2021.
[195] *United States National Action Plan on Women, Peace, and Security* (Washington, DC: White House, 2016), 3.
[196] *Women, Peace, and Security Strategic Framework and Implementation Plan*, 9.
[197] William M. (Mac) Thornberry National Defense Authorization Act for Fiscal Year 2021.
[198] Kelly McCoy, "The Road to Multi-Domain Battle: An Origin Story," Modern War Institute at West Point, 27 October 2017.

such efforts at the GCC level will mean that GCC commanders will become their organizations' executive agents, advised by combatant command-level gender advisors.

DOD Directive 5101.1 defines a DOD executive agent as: "The head of a DoD Component to whom the Secretary of Defense or Deputy Secretary of Defense has assigned specific responsibilities, functions, and authorities to provide defined levels of support for operational missions, or administrative or other designated activities that involve two or more of the DoD Components."[199] Without joint-level executive agents, the third challenge in GCCs' gender perspective implementation may be a lack of direction and focus.

Finally, to succeed at incorporating gender perspective, this research assumes that it is best to follow an orderly change process, such as Kotter's eight-step framework.

Discussion

The WPS SFIP outlines three defense objectives, but only the first is within the scope of this research. Objective 1 is that the DOD should exemplify a diverse organization that allows for women's meaningful participation across the development, management, and employment of the Joint Force.[200] According to Ambassador Melanne Verveer, executive director for the Georgetown Institute for Women, Peace, and Security, in her foreword to *Women and Gender Perspectives in the Military: An International Comparison*, "The promise and potential of United Nations Security Council Resolution (UNSCR) 1325 remains unfulfilled in the field of military operations, which remains a largely male-dominated arena that continues to be held back—in both effectiveness and equal opportunity—by cultural, bureaucratic, and resource barriers."[201]

If DOD is going to implement a gender perspective per the SFIP, it must first look inward. This does not mean forgetting allied, partnered, or adversarial states in a given area of responsibility. Instead, it means improving internal U.S. institutions and organizations first, such as GCCs, to then be able to support external entities more effectively while modeling gender perspective principles abroad. Looking inwardly for DOD translates to GCCs' examining whether and how they can implement the SFIP's gender perspective guidance. Research indicates the benefits may outweigh the costs.[202]

Robert Egnell and Mayesha Alam offer, in their book *Women and Gender Perspectives in the Military: An International Comparison*, that adding gender perspective can transform the traditional military paradigm by looking at all aspects of

---

[199] *Department of Defense Directive 5101.1, Executive Agent* (Washington, DC: U.S. Department of Defense, 3 September 2002), 2.
[200] *Women, Peace, and Security Strategic Framework and Implementation Plan.*
[201] Verveer, foreword to *Women and Gender Perspectives in the Military*, vii.
[202] Egnell and Alam, "Introduction," in *Women and Gender Perspectives in the Military*, 8.

a conflict through varied gender perspectives.[203] Males and females may sometimes look at situations differently, which can expand the aperture of insight, which can therefore improve the overall understanding of a mission, for example.[204]

To implement the WPS SFIP and exploit the advantages that different gender perspectives can offer, this research recommends GCCs follow the eight-step framework for organizational culture change found in John Kotter's book *Leading Change*. It further describes eight errors leaders must avoid in the process. This chapter uses Kotter's steps to guide tailored, embedded suggestions that GCC leaders might follow to incorporate gender perspective within their organizations. It also discusses errors Kotter recommends such leaders avoid.

Kotter defines culture as the "norms of behavior and shared values among a group of people."[205] Because this research assumes change was ordered to address a lack of gender perspective throughout the DOD, GCC leaders must therefore oversee efforts to incorporate it. Here, it should be noted that Kotter distinguishes between *leading* and *managing* change as different processes, and that "management is a set of processes that can keep a complicated system of people and technology running smoothly." He notes that its most important elements "include planning, budgeting, organizing, staffing, controlling, and problem solving."[206] He defines leadership, however, as "a set of processes that creates organizations in the first place or adapts them to significantly changing circumstances." Additionally, "leadership defines what the future should look like, aligns people with that vision, and inspires them to make it happen despite the obstacles."[207]

Understanding the differences between management and leadership is critical, according to Kotter. He asserts that change can be attributed to leadership in 70 to 90 percent of the time and to management only 10 to 30 percent of the time.[208] Therefore, it is important to keep the differences between leadership and management in mind when implementing an organizational culture change toward gender perspective inside the GCCs. Change still requires management, but more importantly, it must start with leadership at all levels to support gender perspective and the benefits it provides the force in a complex environment

Applying Kotter's eight-stage process for leading successful change, the eight steps are:
    1. establishing a sense of urgency
    2. creating the guiding coalition
    3. developing a vision and strategy
    4. communicating the change vision

---

[203] Egnell and Alam, "Introduction," 8.
[204] "Men and Women Explore the Visual World Differently," *ScienceDaily*, 30 November 2012.
[205] John P. Kotter, *Leading Change* (Boston: Harvard Business School Press, 2012), 156.
[206] Kotter, *Leading Change*, 28.
[207] Kotter, *Leading Change*, 28.
[208] Kotter, *Leading Change*, 28.

5. empowering broad-based action
6. generating short-term wins
7. consolidating gains and producing more change, and
8. anchoring new approaches in the culture.[209]

His theory asserts that to successfully incorporate an organizational culture change, such as a gender perspective, requires moving sequentially through all eight steps. He notes that missing a step almost always creates problems.[210] Each step is addressed below, as it applies in GCCs.

Step 1: Establish a Sense of Urgency

According to Kotter, the first stage for leading an organizational culture change is establishing a sense of urgency.[211] DOD has already established one for incorporating gender perspective by publishing its WPS SFIP.[212] A powerful affirmation of and aid to dialogue is incorporating the cause—in this case, gender perspective—into doctrine, as DOD has done with the SFIP. Such documentation communicates validity for gender perspective's importance to GCCs' servicemen and women alike. As Kotter notes, changing doctrine is not a quick solution but an early, essential step toward change.[213] However, just putting the words *gender perspective* onto paper is not enough; leaders must communicate its intent.[214]

In the GCC context, leaders at all levels must explain why gender perspective is essential and how team members can support the mission to incorporate it. The new mandate should trigger dialogue first among senior leaders, not only at the enterprise level but also at the GCC level, about how to resource any changes. U.S. Army War College research professor Dr. Leonard Wong's research states that one must establish events or programs dedicated to an issue to get people talking about that topic.[215] A conference or other discussion forum or working group could involve defining WPS goals and how best to incorporate them into the force.

However, according to Kotter, an early error in leading a change such as gender perspective's incorporation would be for leaders to allow too much complacency within the organization's status quo or to not establish a sense of urgency that change is needed before deciding to implement the new concept.[216] In other words, simply educating people that gender perspective is essential will not bring change. Leaders

---

[209] Kotter, *Leading Change*, 23.
[210] Kotter, *Leading Change*, 25.
[211] Kotter, *Leading Change*, 37.
[212] *Women, Peace, and Security Strategic Framework and Implementation Plan.*
[213] Kotter, *Leading Change*, 36.
[214] Liam Saville, "Communicating Intent: The Importance of Communicating Intent When Implementing Change," Medium.com, 27 September 2019.
[215] Leonard Wong, "Changing the Army's Culture of Cultural Change," Strategic Studies Institute, 16 May 2014.
[216] Kotter, *Leading Change*, 4.

must be intentional about incorporating it throughout their subordinate teams and must empower others to help lead a change with what Kotter calls a "guiding coalition."[217]

Step 2: Create a Guiding Coalition

Once a sense of urgency is established, Kotter recommends creating a guiding coalition.[218] Similar to how an executive agent is needed at the enterprise level, a coalition at the GCC level is also a key aspect of carrying out a culture change. Wong postulates that to change an organization's culture, one must first establish why such change deserves everyone's attention and respect in the organization.[219]

A possible attention-getting tool could be a written gender perspective primer, developed for and distributed to organizational leaders to educate them about gender perspective and how to communicate the concept to their organizations. Such a document would highlight the challenges and benefits of incorporating a gender perspective. For example, a primer or other materials could note that a gender perspective's benefits are consistent with lines of effort in the 2018 *National Defense Strategy* (NDS), which aspires for DOD to strengthen alliances and attract new partners (that is, modeling gender perspective may enhance DOD's soft power with certain allies and partners, particularly in Europe, who enforce gender equality) and to reform the department for greater performance.[220] According to Ambassador Verveer's foreword in *Women and Gender Perspectives in the Military*, "the prevention of conflict, the protection of human rights, and the promotion of peace and security worldwide cannot be achieved without the full and equal participation of women."[221]

The WPS SFIP is also consistent with the NDS's lines of effort, which articulate the benefits of a more diverse fighting force, an improved commitment to human rights and women's empowerment, and effective strategies to abate risks.[222] To demonstrate the relatively low numbers of women in GCCs, leaders may have to focus their organizations' attention on evidence.[223] For example, according to the U.S. Census Bureau, females comprise 50.8 percent of the U.S. population.[224] However, the proportion of women volunteering for service in the U.S. military hovers at around 18.8 percent.[225] Thus, female percentages within the DOD do not yet reflect American society. Within GCCs, the average percentage of women is even smaller: only 12

---

[217] Kotter, *Leading Change*, 51.
[218] Kotter, *Leading Change*, 51.
[219] Wong, "Changing the Army's Culture of Cultural Change."
[220] Magnea Marinosdottir and Rosa Erlingsdottir, "This Is Why Iceland Ranks First for Gender Equality," World Economic Forum, 1 November 2017; and *Summary of the 2018 National Defense Strategy of the United States of America*.
[221] Verveer, foreword to *Women and Gender Perspectives in the Military*, vii.
[222] *Women, Peace, and Security Strategic Framework and Implementation Plan*, 10.
[223] Serota email message.
[224] "Quick Facts, United States," U.S. Census Bureau.
[225] David Vergun, "Service Personnel Chiefs Discuss Diversity in the Military," U.S. Department of Defense, 10 December 2019.

percent.[226] Therefore, getting GCC leaders to attend to or respect the need for a change toward a greater gender perspective within their organizations could prove challenging, particularly among many other competing initiatives and priorities.

To meet such a challenge, a change coalition must have the proven leadership and experience to drive reform.[227] Putting together such a coalition of people to champion a change effort begins with leaders who know which characteristics to look for in any coalition they form to enhance the prospect of achieving the desired outcomes. To create a coalition that supports the sought-after gender perspective inclusion within a GCC, the proponent must not simply rely on senior individuals to help champion change but must also employ those whom Kotter says have "a commitment to improved performance" and who can act as avid proponents.[228] Kotter identifies four key characteristics that the right affiliates should have: position power, expertise, credibility, and leadership.[229] A description of each follows.

- *Position power.* This refers to the stature of coalition members. Ideal coalition members will not only welcome a gender perspective but will also have enough power or influence within the GCC to impede cynics from blocking progress.[230]
- *Expertise.* Coalition members should have knowledge or experience germane to the change effort.[231] Gender advisors may offer initial expertise; however, commanders, staff officers, and senior enlisted members can all train for the requisite knowledge to lead change.
- *Credibility.* Research shows that successful coalition members should have positive reputations within their organizations to gain trust, respect, and others' buy-in for a new idea, which encourages subordinates or other colleagues to take them seriously.[232]
- *Leadership.* GCC commanders' primary staffs could be ideal initial coalition members, followed by others who are qualified to advocate for gender perspective at their levels. Having both empathy and emotional intelligence could enhance all guiding coalition members' ability to gain others' trust and respect. According to Dr. Jean Decety of the University of Chicago, "Empathy consists of both affective and cognitive components and . . . the capacity to adopt the perspective of the other [as] a

---

[226] Serota, email message.
[227] Kotter, Leading Change, 57.
[228] Kotter, Leading Change, 6.
[229] Kotter, Leading Change, 6.
[230] Kotter, Leading Change, 6, 59.
[231] Kotter, Leading Change, 6, 59.
[232] Kotter, Leading Change, 6, 59.

key aspect of human empathy."[233] University of New Hampshire professor John D. Mayer and Yale University professor Peter Salovey add that "emotional intelligence is a type of social intelligence that involves the ability to monitor one's own and others' emotions, to discriminate among them, and to use the information to guide one's thinking and actions."[234]

Kotter's caution for this step involves failing to create a sufficiently powerful, qualified guiding coalition. Some coalitions take decades to get established for action. For example, the Women's International League for Peace and Freedom has worked toward improving women's rights for almost 100 years. However, in 2000 it finally established the PeaceWomen Programme to guarantee that women's rights and participation are acknowledged internationally.[235]

Step 3: Developing a Vision and a Strategy

Kotter notes that "vision plays a key role in producing useful change by helping to direct, align, and inspire actions on the part of large numbers of people."[236] It describes what an organization aspires to be, which helps drive change. The 2021 NDAA states that DOD must implement the WPS SFIP by 30 September 2025. According to Kotter, having a vision statement should aid in meeting that established timeline.[237] Additionally, a vision statement can promote change by providing direction and focus on a common goal, which then lends itself to strategy development. Not having a vision or a strategy are reasons why change efforts can fail.[238]

The Army Vision, Army leadership's concept for the Army of 2028, is a two-page document that states the Army must have a "clear and coherent vision" to accomplish its mission.[239] This is no less applicable for incorporating a gender perspective into a GCC environment. A vision statement on what success will look like for a GCC is required before any strategy to achieve change can begin.

A possible error within this step is "underestimating the power of vision," according to Kotter. An organization can build a strong coalition for change; however, a

---

[233] Kotter, *Leading Change*, 6, 59; and Jean Decety, "Perspective Taking as the Royal Avenue to Empathy," in *Other Minds: How Humans Bridge the Divide Between Self and Others*, ed. Bertram F. Malle and Sara D. Hodges (New York: Guilford Press, 2005), 143–45.
[234] John D. Mayer and Peter Salovey, "The Intelligence of Emotional Intelligence," *Science Direct* 17, no. 4 (October–December 1993): 433–42, https://doi.org/10.1016/0160-2896(93)90010-3.
[235] *Women, Peace and Security National Action Plan Development Toolkit* (New York: PeaceWomen of Women's International League for Peace and Freedom, 2013), 4.
[236] Kotter, *Leading Change*, 7–8.
[237] William M. (Mac) Thornberry National Defense Authorization Act for Fiscal Year 2021, Pub. L. No. 116-283 (2020); and Kotter, *Leading Change*, 8–9.
[238] Mark Lipton, "Demystifying the Development of an Organizational Vision," *Sloan Management Review* 37, no. 4 (Summer 1996): 83–92.
[239] Mark T. Esper and Mark A. Milley, *The Army Vision* (Washington, DC: U.S. Department of the Army, 2018), 1.

poorly written or insufficient vision can lead the coalition in the wrong direction. As Kotter notes, without a vision, a change effort can be overcome by incoherent or even unrelated tasks that do not lead to the desired outcome.[240] In other words, without a vision, gender perspective is a verbal ambition that perhaps no one will implement.

Step 4: Communicating the Change Vision

Once goals and a vision are established, Kotter recommends "communicating the change vision."[241] He notes there are seven effective elements to communicating a vision effectively: 1) simplicity, 2) analogy, 3) multiple forums, 4) repetition, 5) leading by example, 6) explaining inconsistencies, and 7) establishing two-way communication.[242] A discussion of each element's application in a GCC follows.

To start, a simple vision must be clear enough for every hearer or reader to understand it. One such clear example is a statement by the NATO secretary general's special representative for WPS, Clare Hutchinson, that NATO's "vision of security must be anchored to the inclusion of women, the adoption of a gender perspective in all activities, and in upholding the highest standards of behavior."[243]

The second element is providing an analogy that paints a picture for listeners, viewers, or readers.[244] For example, a leader might describe a failure to employ women as force multipliers downrange as a choice to "row a boat with only one oar and go in circles" rather than to make progress. Memorable metaphors and other analogies spring to mind when heard or seen often as reminders of the need to use all resources available, just as in the Second World War, when images of Rosie the Riveter flexing her muscles communicated women's capabilities to keep U.S. industry humming.[245]

The third element is using multiple forums to get out the word (that is, the vision statement) to the GCCs' members.[246] For example, top-level meetings, recreational events, or even unit formations can give leaders and others a platform to model some aspect of gender perspective. For example, an article in *Military Times* described an example of gender perspective exhibited by a male soldier toward Captain Kristen Griest, one of the Army's first female infantry officers and one of the first women to complete Army Ranger School, before women were accepted into the infantry. Once policy changed to allow her to command a company within the 505th Infantry Regiment, 3d Brigade Combat Team, she expected resistance to her arrival.[247] However, a

---

[240] Kotter, *Leading Change*, 7–8.
[241] Kotter, *Leading Change*, 85–86.
[242] Kotter, *Leading Change*, 90–92.
[243] "Twenty Years on, NATO's Commitment to Women, Peace, and Security Is Stronger than Ever," North Atlantic Treaty Organization, 3 November 2020.
[244] Kotter, *Leading Change*, 90–92.
[245] "Rosie the Riveter," History.com, 9 February 2021.
[246] Kotter, *Leading Change*, 90.
[247] Meghann Myers, "First Female Ranger Grads Open Up about the Aftermath and Joining the Infantry," *Army Times*, 13 March 2018.

former Captain's Course colleague of hers greeted her enthusiastically and respectfully in front of their formation instead. He looked past her gender and welcomed her as a member of one team.

The fourth element is repetition for retention to keep standards as readily in mind as, for example, the five-paragraph operations order outline, which all officers memorize.[248] Likewise, a vision statement must be repeated until ingrained. Kotter warns that "undercommunicating the vision" happens "by a factor of 10 (or 100 or even 1,000)," which can hinder stakeholders' momentum for change.[249] People will follow the vision of a leader they believe in if that leader convincingly communicates the benefits of a change early and often in various ways. Nevertheless, regardless of how unhappy servicemembers may be, they will not sacrifice the comfort of the known unless they accept the change being communicated.[250] Therefore, consistent communication about gender perspective's benefits will be a critical component in gaining support for it.

Conversely, Kotter notes three patterns of ineffective communication that leaders encourage when servicemembers are comfortable with present circumstances. The first pattern is when leaders state a strong vision but do not sell it to their subordinates.[251] In effect, such leaders abandon communicating the vision's importance and are later surprised at not getting results. To avoid this error with gender perspective and to spark real change, GCC leaders should develop "marketing plans" for disseminating their vision on gender perspective in their mission set. Army Regulation 601-208 states that a marketing plan "identifies relevant prospect and influencer audiences, directing the appropriate brand communication to that audience within the appropriate media at the appropriate time."[252]

A second pattern of ineffective communication is when leaders are the only figures articulating the vision while subordinates remain indifferent.[253] Leaders need proponents throughout their GCCs who support and amplify the vision. A third ineffective communication pattern to avoid is permitting leaders to communicate nonsupport for an enacted initiative.[254] For example, having an influential person stand in the back of the room with arms crossed or making negative comments during a town hall, for example, can derail what ordinarily might have been an opportunity for neutral or positive dialogue. Change leaders must identify such negative stakeholders early on to stymie their effects.

The fifth element for collectively communicating a vision is to lead by example as

---

[248] Kotter, *Leading Change*, 90; and *Train to Win in a Complex World*, Field Manual 7-0 (Washington, DC: U.S. Department of the Army, 2016).
[249] Kotter, *Leading Change*, 9.
[250] Kotter, *Leading Change*, 9–10.
[251] Kotter, *Leading Change*, 9–10.
[252] *Personnel Procurement: The Army Brand and Marketing Program*, Army Regulation 601-208 (Washington, DC: U.S. Department of the Army, 2013), 10.
[253] Kotter, *Leading Change*, 9.
[254] Kotter, *Leading Change*, 9.

a consistent proponent of the vision.[255] This amounts to the leadership maxim "never ask someone to do something that you yourself are not willing to do." Being a consistent proponent means always moving in the direction of the vision by advocating for gender perspective and how it can improve one's organization for the better.

The sixth element is explaining inconsistencies that may undermine the credibility of communication.[256] For example, the U.S. Army is very consistent in its communication of Army values, which include respect, honor, and integrity; however, 3,219 reports of sexual assaults allegedly committed by soldiers were filed by fellow servicemembers in 2019 alone.[257] Despite espousing Army values with consistency, such assaults undermine Army values' credibility.

Finally, two-way communication must be established between coalition members and the soldiers they hope to influence.[258] GCC leaders must address both positive and negative feedback about changing to a gender perspective. Such communication must allow the lowest member of the organization ownership and a voice, and must allow ideas to emanate from any level.

Step 5: Empowering a Broad Base of People to Take Action

Kotter's fifth step is to empower a variety of team members for action.[259] The best way to do this is to discover and remove barriers to change. For example, in the article about Captain Griest cited earlier, an assumption may be that other male soldiers wanted to welcome her but may have feared how their comrades might react. If just one respected person takes the initiative to model a behavior, as her former classmate did, such behavior can empower others to follow suit.

Kotter offers the following options for doing so once the leadership communicates "a sensible vision." First, all structures that are set up to further the vision must be compatible with it. Second, soldiers must have access to, and time granted for, any training they may need. Third, information and personnel systems must align with the vision. Finally, leaders and their coalition members must confront supervisors who undercut needed changes.[260] All the options above can guide GCC leaders in implementing gender perspective within their organizations.

"Permitting obstacles to block the new vision" is an error Kotter warns of, asserting that "new initiatives fail far too often when employees, even though they embrace

---

[255] Kotter, *Leading Change*, 90.
[256] Kotter, *Leading Change*, 90.
[257] *Personnel-General: Army Profession and Leadership Policy*, Army Regulation 600-100 (Washington, DC: U.S. Department of the Army, 2017), 31; and Patricia Kime, "Sexual Assault Reports, Harassment Complaints Rise in the US Military," Military.com, 30 April 2020.
[258] Kotter, *Leading Change*, 90.
[259] Kotter, *Leading Change*, 106.
[260] Kotter, *Leading Change*, 115–19.

a new vision, feel disempowered by huge obstacles in their paths." Such obstacles do not necessarily need to be real. As shown in the formation example above, Griest's expectation of how she would be received was a perception, not a reality. Kotter states that sometimes the obstacle may only be conceptual. The challenge is in convincing people that no external barriers exist, as Griest's colleague did with his simple, respectful action.[261]

Asking soldiers for their input is another simple way to not only identify real or perceived barriers but also to garner stakeholder support for gender perspective-related change or to do what Kotter calls "empowering people to effect change."[262] For example, the Army's 18th Airborne Corps attempted to eliminate sexual assault in its ranks by asking soldiers to submit ideas on how to succeed. Seven of the ideas selected from 41 total submissions are now being implemented.[263]

To get such buy-in and feedback, GCC leaders can schedule town hall meetings with their organizations' members at all levels to explain the benefits and the importance of gender perspective to their operations. For example, during Lieutenant General Darryl A. Williams's first town hall meeting as the superintendent of the U.S. Military Academy at West Point, he stated that everyone at the school should "treat each other with dignity and respect" and that "at the end of the day, it is about treating people the way you want to be treated."[264] Moreover, MIT lecturer Douglas A. Ready posits that leaders commit to both a communication and a listening campaign so that everyone is aware and understands how to contribute to the organizational mission.[265]

Beyond these tools, research shows that leaders can exploit approaches to change by showcasing examples of how gender perspective improved other organizations' outcomes. For example, Swedish Armed Forces deploying to Afghanistan designated selected personnel as gender focal points (GFPs) within each platoon or staff section prior to arrival. GFP duties included dealing with gender issues in operations. These individuals were instrumental to gender advisors by providing reports that helped with developing gender-issue training. This ultimately improved the Swedish Armed Forces' mission success.[266] GCC commanders could emulate GFP methods to incorporate gender perspective-related initiatives into U.S. operations.

Gender perspective exemplars need not be women. Research shows that gender

---

[261] Kotter, *Leading Change*, 10.
[262] Kotter, *Leading Change*, 115–19.
[263] Caitlin M. Kenney, "Seven Soldiers to Present Their Ideas on Improving the Army's Program Targeting Sexual Assault," *Stars and Stripes*, 18 February 2021.
[264] Brandon O'Connor, "Superintendent Outlines Vision for USMA during Town Hall," *Pointer View*, 23 August 2018.
[265] Douglas A. Ready, "4 Things Successful Change Leaders Do Well," *Harvard Business Review*, 28 January 2016.
[266] Robert Egnell, "Sweden's Implementation of a Gender Perspective: Cutting Edge but Momentum Lost," in *Women and Gender Perspectives in the Military*, 49.

perspective-informed and supportive men can be just as effective at incorporating changes that lead to a gender perspective-infused organizational culture as women with similar skills.[267] Lieutenant Colonel Scott Stephens, for example, became a leading voice opposing sexual harassment and assault. He and other soldiers wrote *Athena Thriving*, an educational resource for leaders to better understand the gender-based issues women in the Army face.[268] GCC commanders can encourage or designate such motivated team members to champion gender perspective within the organization to help educate others via train-the-trainer instruction.

Step 6: Generating Short-term Wins

Step 6 in Kotter's process for leading successful change is "generating short-term wins."[269] Assigning professional, fulltime gender advisors and WPS analysts to GCCs is a short-term win. Such experts can focus not only on implementation but also on educating leaders and soldiers on why gender perspective is important and where opportunities for its implementation exist.

Ignoring short-term wins is a mistake, according to Kotter. If the intended solution to a problem will take a considerable amount of time, leaders must score incremental initiatives quickly to support longer-term goals. When change fails, Kotter posits no systematic effort to guarantee unambiguous wins occurred within the first 6 to 18 months of attempted implementation.[270] Leaders cannot expect to see positive results in the long-term if they do not first show them in the short-term.

Step 7: Consolidating Gains and Producing More Change

In Kotter's seventh step, "consolidating gains and producing even more change," the established gender perspective coalition would build on its credibility from earlier stages to encourage others within the GCC to join their effort.[271] At this point in the process, change should begin to take shape as more servicemembers see a gender perspective's benefits to the organization.

However, "declaring victory too soon" is a mistake, notes Kotter.[272] Leaders must accept that real culture change can and does take a long time. He warns that new approaches are delicate and are easily undone, unless or until they get rooted in the organizational culture.[273] Day-to-day requirements for GCCs are vast and can potentially overshadow any drive for change that may take years to achieve. Leaders must therefore focus on long-term, lasting change.

---

[267] Egnell and Alam, "Introduction," 14.
[268] Haley Britzky, "This Army Lieutenant Colonel Has Built a Playbook to Kill the 'Cancer' of Sexual Assault in the Ranks," *Task & Purpose*, 1 March 2021.
[269] Kotter, *Leading Change*, 117.
[270] Kotter, *Leading Change*, 11–12.
[271] Kotter, *Leading Change*, 131.
[272] Kotter, *Leading Change*, 13.
[273] Kotter, *Leading Change*, 14.

Step 8: Institutionalizing New Approaches in the Culture

Kotter describes his final step as "anchoring new approaches in the culture."[274] This may be the most difficult one of all. In his book *Who Says Elephants Can't Dance? Inside IBM's Historic Turnaround*, Louis V. Gerstner Jr. observed that "changing the . . . behavior of . . . thousands of people is very . . . hard. . . . Business schools don't teach you how to do it, and you can't mandate . . . or engineer it."[275] Changes to norms and values often come at the end of a change process.[276]

Final errors happen when organizations desert change before it gets ingrained. To avoid this, GCC leaders must show their forces how a shift in culture improved their organization, rather than expecting soldiers to discern this alone. Leaders need to articulate a gender perspective's positive outcomes and recognize helpful behaviors and attitudes that show that the organization embodied positive change.[277] As GCC commanders depart, it is also critical that outgoing leaders inform incoming ones of the hard-won changes their team implemented.

John Kotter's Leading Change model can offer GCCs a structured and disciplined framework for incorporating a gender perspective into their organizations' cultures. As in all large organizations with longstanding traditional methods of operating, change is challenging and therefore requires a disciplined approach, such as that offered by Kotter's eight-step change process. Before their organizational cultures can change permanently to ones that incorporate gender perspective, GCCs need to establish or use a clear vision and strategy, create an internal coalition, use clear and persistent communication, empower subordinate agents for change, and achieve visible short-term wins and gains. In the GCC context, as in others, errors along the path to change must also be avoided, such as poor leadership, unclear objectives, inconsistent communications, or a lack of will or progress.

U.S. GCC leaders and their teams are among the best the DOD can offer and are capable of meeting worldwide challenges. Incorporating the organizational culture change that gender perspective requires is an internal challenge that GCCs are well prepared to overcome. Doing so is not only mandated, but it is also welcome as an opportunity to operate better from a diversity standpoint and to close a longstanding "say-do" gap between U.S. rhetoric about gender equality and how America's most visible global military ambassadors actually look and operate abroad.

---

[274] Kotter, *Leading Change*, 131.
[275] Louis V. Gerstner Jr., *Who Says Elephants Can't Dance?: Inside IBM's Historic Turnaround* (New York: Harper Collins, 2003), 378.
[276] Kotter, *Leading Change*, 157.
[277] Kotter, *Leading Change*, 14–15.

Chapter 5
Conclusion

by Colonel Douglas Winton, USA*

The resounding theme of the research and analysis conducted by the authors of the preceding chapters revolved around the value of a gender perspective, one of the global WPS principles. However, it is only one of seven principles that the WPS SFIP outlines: the participation of women in peace and security; the protection of women and girls from violence; the inclusion of women in conflict prevention; equal access to relief and recovery before, during, and after conflict and crisis; the protection of human rights; equal application of the rule of law; and incorporating a gender perspective into peace and security efforts.[278]

These principles are global, however, there are national lines of effort and departmental equities that support the WPS principles within U.S. government and military bodies. WPS is a complex agenda that requires an up-and-out and a down-and-in approach to successfully implement it as directed within the 2021 NDAA within the DOD.

Although the focus of these chapters centers around the utility of a gender perspective, there remain myriad topics to address and research. Moreover, it is essential to state that time is of the essence; September 2025 is not far distant.

So, what is missing from this research? The authors posit more than can be covered in a small research project paper. Future Army War College students must continue to confront the subject to assist with the WPS strategic research to enhance successful implementation. Additionally, during the research it became apparent that the implementation and training of WPS are somewhat disjointed and will require future dialogue. This chapter addresses recommended areas for future follow-on research. The breadth and depth of WPS are vast, and it is to be hoped that future students and leaders throughout the DOD take this topic seriously.

As previously discussed, the authors recognized that to address the entire WPS agenda was impossible and so strategically chose gender perspective because of the topic's complex nature. Therefore, the first item missing from this research dealt with the concept of a disjointed effort on training and implementation. When looking at the GCCs through the Unified Command Plan lens it becomes clear that they are all addressing WPS differently. To implement WPS, there must be a foundation established within the DOD outlining the basic implementation guidelines, which is not

---

* The views expressed in this chapter are solely those of the author. They do not necessarily reflect the opinion of Marine Corps University, the U.S. Marine Corps, the U.S. Navy, the U.S. Army, U.S. Army War College, the U.S. Air Force, or the U.S. government.

[278] *Women, Peace, and Security Strategic Framework and Implementation Plan* (Washington, DC: U.S. Department of Defense, 2020), 9.

currently the case. All GCCs are implementing WPS based on their specific missions within their regions of influence. Not having a program of instruction directed by DOD leaves each GCC to implement as they see fit. The problem with this is that when an individual leaves one command for another command, the education gained may not work effectively within the gaining unit. There is a reason why training institutions utilize a program of instruction; it ensures everyone training on a subject gets the same foundational skills needed throughout the Service. GCCs' implementation of WPS should be contextualized for the issues in their specific regions, but that contextualization should be based on a common set of DOD standards and practices as detailed in a coherent program of instruction.

Additionally, GCCs have hired gender advisors to make recommendations to the command teams and leadership within the GCCs. However, they are busy frantically building training packages intending to educate the force on WPS. Again, if each GCC is working to implement and train in a vacuum, it will create gaps that future commanders must overcome to implement WPS across the force successfully.

The authors of the three preceding chapters each defined *gender perspective* slightly differently, and each had a slightly different slant on how to address the problem through their research. In the same way, three different authors writing on the same WPS principle would likely come to three very different conclusions. One can only assume a similar effect is at play among the different DOD commands implementing WPS as they each understand it and achieving different outcomes.

Another reason training and implementation is disjointed is the lack of an executive agent for WPS. *DOD Directive 5101.1, DOD Executive Agent* defines an executive agent as "the Head of a DoD Component to whom the Secretary of Defense or the Deputy Secretary of Defense has assigned specific responsibilities, functions, and authorities to provide defined levels of support for operational missions, or administrative or other designated activities that involve two or more of the DoD Components."[279] Currently, the DOD is the executive agent for all Services to implement WPS, which the authors claim will potentially overwhelm the secretary of defense and their staff. The DOD must consider assigning an executive agent as defined in *DOD Directive 5101.1* to provide direction on training and implementation instructions. Without an executive agent, each Service is left to implement and develop training on its own, which will lead to a less than adequate implementation of WPS by September 2025.

While these chapters briefly touched on effective implementation, they did not address what effective implementation of WPS looks like in the future. In the military, leaders are always asking what success looks like after implementation. The DOD must ask a similar question about WPS and its global, national lines of effort and departmental equities. Is the implementation of all three the only way to succeed, or

---

[279] *Department of Defense Directive 5101.1, DOD Executive Agent* (Washington, DC: U.S. Department of Defense, 3 September 2002), 2.

can there be parts of each implemented for success? This is another area that was not explicitly called out in any of the three chapters.

Recommendations for further research within the WPS agenda are a strategic imperative, and the authors make five suggestions with potentially wide-ranging impacts throughout the global, national, and departmental areas of WPS. DOD must concentrate on diversity and inclusion and on recruitment and retention; to seek and support the meaningful participation of women in the military decision-making process, according to the first national line of effort; and encourage and support the participation of women in peace and security.[280] Finally, DOD must address how it operationalizes WPS throughout the force.

The list of topics above is by no means all-inclusive but is more of a starting point for future students to consider. These chapters addressed diversity and inclusion, recruitment, and retention. The authors suggest that improving these principles within the DOD will positively impact WPS across the defense enterprise; however, more specific research is warranted.

Improving the participation of women in the military decisionmaking process speaks to the potential contributions of a gender perspective; however, explicitly researching this area may lend itself to improvements throughout the DOD. The improvements referred to are a complete and thorough course of action, which leads to a more lethal force through the inclusion of women in the process.

Future researchers should also look at the participation of women in peace and security from the WPS principles. The recommendations all speak to how to better include women into the decision process, which ultimately supports the operationalization of WPS within the DOD. The research revealed that the U.S. population is greater than 50 percent female, and if women have a different perspective than men, why wouldn't the DOD want to include them?[281] Again, this is an important research topic for students in next year's academic cohort to address.

Finally, more work must be devoted to including WPS within the military education system, including WPS training for noncommissioned officers and officer professional military education courses, such as the Basic Leadership Course and Basic Officer Leadership Course. It will help set the stage for successful implementation within the DOD. Additionally, it must go even further. It must inculcate WPS into Basic Combat Training, Advanced Individual Training, and officer-producing courses including Officer Candidate School, Reserve Officer Training Corps, and the U.S. Military Academy. With little time left to implement according to the 2021 NDAA, the time to further research how to inculcate WPS into initial entry training and PME courses is now. This will be a significant research project for a future group to analyze and offer potential courses of action.

---

[280] *Women, Peace, and Security Strategic Framework and Implementation Plan*, 9.
[281] "Quick Facts, United States," U.S. Census Bureau.

How does the DOD further expand WPS within the Services? Each Service and component is currently left to do it alone and has received little guidance or direction. Further research is warranted in the coming academic years across the PME continuum and is a great resource to further the WPS agenda beyond its current state. The authors hope that the staff at the Army War College find a sponsor and put together a robust integrated research group to ensure this topic garners the attention it deserves.

PART 2

Gender Neutrality

Gender-Neutral Physical Fitness Tests and the Integration of Women in Combat Arms Occupations

by Second Lieutenant Elizavetta Fursova*

Introduction

The Combat Exclusion Policy was repealed in 2013, which opened all occupations within the military—to include combat arms—to women.[1] Since then, however, women have faced a number of barriers to integration into these previously closed positions in the U.S. military. Specifically, even as the military shifted toward gender-neutral physical fitness standards, long-standing social biases and physiological differences between men and women have contributed to contentions regarding the integration of women into combat arms specialties. Policymakers moved toward gender-neutral occupational standards as a more systematic way to determine who would be qualified to fill positions with physically demanding tasks.[2] Where gender previously served as a proxy screener for physical ability in jobs, the Services were now faced with "the possible inclusion of larger numbers of personnel who could not meet the physical demands," leading to the development of gender-neutral fitness standards.[3] Arguably, a gender-neutral physical fitness assessment would allow the military to make "the best possible use of talents and capabilities of women who sought combat positions."[4]

Societal expectations and biological differences that lead to varying physical abilities previously motivated the exclusion of women from combat arms branches and roles.[5] Similarly, debates on the validity of integration of women into combat arms branches stem largely from the "band of brothers" myth, which is a "generally accepted and consistent set of narratives linked to all-male units, male bonding, cour-

---

* The views expressed in this chapter are solely those of the author. They do not necessarily reflect the opinion of Marine Corps University, the U.S. Marine Corps, the U.S. Navy, the U.S. Army, U.S. Army War College, the U.S. Air Force, or the U.S. government.
[1] Notice to Congress of proposed changes in units, assignments, etc. to which female members may be assigned, 10 U.S.C. § 652 (2012).
[2] Chaitra Hardison, Susan Hosek, and Chloe Bird, *A Review of Best Practice Methods*, vol. 1, *Establishing Gender-Neutral Physical Standards for Ground Combat Occupations* (Santa Monica, CA: Rand, 2018), xii.
[3] Various documents published by the DOD, such as *DOD Instruction 1300.28, In-Service Transition for Transgender Service Members* (Washington, DC: Office of the Under Secretary of Defense, 1 October 2016) tend to conflate the terms *sex* and *gender*. While the DOD policy on transgender servicemembers continues to develop, clarification of these terms and their use is necessary for accurate understanding of this research. While *sex* refers to a person's biological characteristics at birth, *gender* refers to a person's self-identification as either male or female regardless of their sex. For the purposes of this paper, further references to *gender* describe an individual's concept of themselves and their self-identification as either male or female. Hardison, Hosek, and Bird, *A Review of Best Practice Methods*, xii.
[4] Ash Carter, "No Exceptions: The Decision to Open All Military Positions to Women," BelferCenter.org, December 2018.
[5] Michele M. Putko, "The Combat Exclusion Policy in the Modern Security Environment," in *Women in Combat Compendium*, ed. Michele M. Putko and Douglas V. Johnson Jr. (Carlisle, PA: U.S. Army War College Strategic Studies Institute, 2008), 27.

age under fire, and the protection of the nation," as well as from the physiological differences between men and women.[6] Because men tend to be faster and stronger than women, men are perceived as "natural warriors."[7] Generally, women have a lower VO2 max, weaker upper body strength, and less muscle mass.[8]

Given these physiological and biological differences, women have struggled to compete with men in combat arms entry screening tests and, as a result, do not experience desired levels of quantitative integration. Although DOD has not provided a set quantitative goal for integration, former secretary of defense James N. Mattis asserted that "there are too few . . . stalwart young ladies . . . charging into this [combat occupations]."[9] Especially in the Marine Corps, where women only comprise eight percent of the force and attrition rates for combat training are high, quantitative integration of women is low.[10] The Commandant of the Marine Corps, General David H. Berger, recognized gender integration as an area of improvement and tasked his subordinates to seek out active-duty female officers to attend the Infantry Officer Course.[11] As such, though gender-neutral physical tests have paved the way for women to participate in the armed forces in previously out of reach ways, there are still concerns about quantitative integration.

Notably, there are certain areas in which women's performance surpasses that of men. Specifically, women generally have greater aerobic capacity and resistance to muscular fatigue, as well as faster recovery following exercise.[12] Most military tests in their current form favor male physiological strengths (upper body strength and anaerobic capacity), resulting in lower women's average scores for physical fitness tests.[13] The military's gradual transition to more gender-neutral standards has thus complicated the conversation around women's service in certain military occupations as new gender-neutral standards favor larger and stronger men.[14]

The physiological and biological differences between men and women also highlight that the current physical test standards compound the performance gap in

---

[6] Megan Mackenzie, *Beyond the Band of Brothers: The US Military and the Myth that Women Can't Fight* (New York: Cambridge University Press, 2015), 14.
[7] Anthony King, "The Female Combat Soldier," *European Journal of International Relations* 22, no. 1 (March 2016): 123, https://doi.org/10.1177/1354066115581909.
[8] VO2 max is defined as the maximum volume of oxygen the body can deliver to the working muscles per minute. The higher the VO2 max, the faster one can expect to run. Alexander Hutchison, "Lactate Threshold and VO2 Max Explained," Active, accessed 5 November 2021.
[9] James N. Mattis, "Remarks by Secretary Mattis at the Virginia Military Institute, Lexington, Virginia" (speech, Virginia Military Institute, Lexington, VA, 25 September 2018), hereafter Mattis remarks.
[10] Emma Moore, "Women in Combat: Five-Year Status Update," Center for a New American Security, 31 March 2020.
[11] Moore, "Women in Combat."
[12] Thomas S. Szayna et al., *Considerations for Integrating Women into Closed Occupations in U.S. Special Operations Forces* (Santa Monica, CA: Rand, 2015), 45.
[13] Szayna et al., *Considerations for Integrating Women in Closed Occupations in U.S. Special Operations Forces*, 49.
[14] Data gathered from 4th Infantry Division Army Combat Fitness Test scores, as described by male Army officer (anonymous), interview with author, 29 January 2021, hereafter 29 January Army officer interview.

gender. And though the military is moving toward more effective measurements of combat-related fitness, the extent to which the current fitness evaluations approximate combat tasks is questionable. As a result, physical standards remain a common way that women are excluded from fully participating in combat arms branches, even with the implementation of gender-neutral physical standards.

Moreover, it is unclear whether the quality of integration under existing gender-neutral physical standards is acceptable to policymakers and women themselves. The U.S. military today is faced with the issue of integrating more women into previously closed all-male occupations, so it is important to identify ways in which certain physical standards stifle the progress and effectiveness of female integration.

Accordingly, the research question examined in this chapter is: *what effect has the adoption of gender-neutral physical standards had on the integration of women in combat arms units?* To consider the question, this research also examines whether implementing gender-neutral standards affects these outcomes that are relevant to female integration. How are changes toward gender-neutral physical standards perceived by the force and how does this influence attitudes toward women in the military? This study sought to determine the ways in which physical standards affect women's ability to integrate into previously closed combat arms branches within the military by assessing the perceptions of the effectiveness of a physical assessment standard on the inclusion and attitudes of units. It may be expected that the perception of a standard's gender neutrality depends on whether that standard has been changed since the repeal of the combat exclusion policy in 2013, the organizational culture of a unit, and the type of gender neutrality one ascribes to. Possible types of gender neutrality could revolve around a single metric used to evaluate both genders or could be more focused on performance outcomes of soldiers conducting a physical fitness assessment.

Current literature on female integration into combat arms branches has not sufficiently considered the perception of physical standards on unit cohesion and attitudes toward women fulfilling combat roles traditionally filled by men. Although women now have access to combat arms occupations in the military, qualitative acceptance of women in these fields varies.[15] Understanding the effect of gender-neutral physical standards on the inclusion, attitudes, and cohesion of servicemembers in combat units would help the military move toward a more effective talent assessment that gives equal opportunity to all soldiers.

My study examines the physical fitness assessments of the Army and the Marine Corps with a specific focus on the Army Ranger School Ranger Assessment Phase (RAP), the Marine Corps Infantry Officers Course (IOC), and the Army Combat Fitness Test (ACFT). RAP has not changed its physical assessment standards since the 1990s and consists of the Ranger Physical Fitness Test, two timed ruck marches, and several

---

[15] Moore, "Women in Combat."

runs.[16] On the other hand, the physical fitness requirements of the IOC changed in 2018, when the Combat Endurance Test (CET) no longer became a graduation requirement.[17] Although little information on the specifics of the CET is available to the general public due to the secretive nature of the event, the test is known to involve several ruck marches under a heavy load (up to 120 pounds).[18] The removal of the CET as a requirement to pass IOC generally coincided with the integration of women into the Marine Corps in 2016. As such, some interpreted the change in IOC physical testing standards as the military lowering standards in order to accommodate this integration.[19] The ACFT—the Army's newest fitness test—replaced the Army Physical Fitness Test (APFT) and evaluates more physical modalities, resulting in a general perception that it is a more comprehensive test of soldier fitness.[20] Unlike the APFT, the ACFT has the same grade scale for women and men. However, the performance gap between sexes on the ACFT has caused concern from servicemembers and policymakers alike, who fear generally poor female performance on the ACFT "could damage some soldiers' professional prospects."[21] This study distinguishes between two types of gender neutrality in relation to physical standards, and situates these definitions in the context of cohesion and attitudes toward women in combat arms.

Literature Review
The Social Dimension

To understand the effect of gender-neutral physical standards on cohesion, attitudes toward, and integration of women in combat arms, it is important to first understand factors that influence cohesion, attitudes, and inclusion. Scholars distinguish between social cohesion and task cohesion. Social cohesion requires the absence of latent social conflict (income inequality, racial or ethnic tensions, and other forms of polarization including gender inequality) and the presence of strong social bonds.[22] Task cohesion is fostered when groups work together to accomplish a given task and a group identifies strongly with the mission of accomplishing this task.[23]

Scholars generally study military or unit cohesion separately from social cohesion and task cohesion. In 1950, Leon Festinger, Kurt Back, and Stanley Schachter developed

---

[16] *Student Information: Ranger School Phases* (Fort Benning, GA: Maneuver Center of Excellence), accessed 5 November 2021.
[17] Shawn Snow, "Passing Combat Endurance Test Is No Longer Required for Infantry Officers," *Marine Corps Times*, 7 February 2018.
[18] Hope Hodge Seck, "Three Women Dropped from Marines' Infantry Officer Course Will Not Reattempt," *Marine Corps Times*, 28 October 2014.
[19] Heather MacDonald, "Women Don't Belong in Combat Units," *Wall Street Journal*, 16 January 2019.
[20] "The Army Combat Fitness Test," Performance Triad, 13 February 2020.
[21] Missy Ryan, "Senators Urge Pentagon to Suspend Implementation of Army's New Fitness Test," *Washington Post*, 20 October 2020.
[22] Ichiro Kawachi and Lisa F. Berkman, "Social Capital, Social Cohesion, and Health," in *Social Epidemiology*, ed. Lisa F. Berkman, Ichiro Kawachi, and Maria Glymour, 2d ed. (New York: Oxford University Press, 2014), 175.
[23] Mark Eys and Albert Carron, "Role Ambiguity, Task Cohesion, and Task Self-Efficacy," *Small Group Research* 32, no. 3 (June 2001): 356–73, https://doi.org/10.1177/104649640103200305.

a new definition of group cohesion that has been used by many researchers since. They defined group cohesion as "the desire of individuals to maintain their affiliation with a group, and this drive is measured by influence and initiative, task competence, and especially like-dislike."[24] The most commonly used definition of social cohesion focuses narrowly on interpersonal attraction, but the definition of requirements for military cohesion incorporates trust and social support built within groups.[25] In other words, military cohesion is centered around interpersonal bonds and the trust that often results from task-related competencies, including perceived or actual physical fitness measured by physical fitness assessments.[26] Moreover, Edward Shils and Morris Janowitz found that "community of experience" such as shared hardship breeds solidarity and thus increases cohesion.[27]

The band of brothers myth reinforces the argument that the adoption of gender-neutral physical standards and integration of women into combat specialties undermines social cohesion. This myth revolves around three key assumptions commonly internalized by warfighters. First, the myth casts male bonding and feelings of "trust, pride, honor, and loyalty" between men as exceptional. Second, it holds exclusively male bonding as an essential element of an orderly and civilized society. Third, it maintains that "all male units are seen as elite as a result of their social bonds and physical superiority."[28] This final assumption rests on the physiological differences between men and women and emphasizes the presumed inability of men to view women as anything other than objects of sexual attraction. Those who oppose female participation in combat operations tend to draw on the presumption that integration of women into such units would create insurmountable tensions and frictions. Those who ascribe to this opinion believe "there is not a woman alive who could contribute enough to one of [a commander's] teams over the long haul to make up for what her presence would do to the trust among his men."[29] The band of brothers myth therefore provides the context in which soldiers may perceive a change in physical standards (toward gender-neutral standards). Physical and mental abilities, accordingly, may not represent the greatest obstacle for women's integration in combat arms branches—social and cultural issues stemming from the band of brothers myth and exacerbated by in-group/out-group dynamics more accurately represent the obstacles women have to face.

---

[24] Xavier Fonseca, Stephan Lukosch, and Frances Brazier, "Social Cohesion Revisited: A New Definition and How to Characterize It," *Innovation: The European Journal of Social Science Research* 32, no. 2 (2019): 233, https://doi.org/10.1080/13511610.2018.1497480.
[25] Szayna et al., *Considerations for Integrating Women into Closed Occupations in the U.S. Special Forces*, 79.
[26] Szayna et al., *Considerations for Integrating Women into Closed Occupations in the U.S. Special Forces*, 80.
[27] Edward Shils and Morris Janowitz, "Cohesion and Disintegration in the Wehrmacht in World War II," *Public Opinion Quarterly* 12, no. 2 (Summer 1948): 287, https://doi.org/10.1086/265951.
[28] Mackenzie, *Beyond the Band of Brothers*, 2–3.
[29] Anna Simons, "Women Can Never 'Belong' in Combat," *Orbis* 44, no. 3 (June 2000): 461, https://doi.org/10.1016/S0030-4387(00)00037-5.

Contrary to some opinions and the band of brothers myth, studies conducted within the armed forces of the United States demonstrate no negative effect on cohesion in mixed-gender units. Such conclusions may suggest that gender-neutral fitness tests will have little impact on cohesion. A Marine Corps study from 2015 as well as a study conducted by the General Accounting Office after the Persian Gulf War both found that gender was not a determinant or component of cohesion and that bonding in mixed-gender units was sometimes even better than unit cohesion in single-gender units.[30] However, the narrative surrounding the Marine Corps study largely misinterpreted the conclusions, and thus undermined the validity and applicability of this study.[31] Critiques of this study center around its overgeneralization of female performance, the failure to establish gender-neutral standards or outline specific pass/fail standards for each group, and a lack of time spent in each group.[32] Current literature largely fails to consider the implications of gender-neutral physical standards on cohesion in mixed-gender units, warranting further study of the topic.

Physical Fitness and Gender
Numerous studies exist that demonstrate the aforementioned physical, biological, physiological, and therefore performance differences between men and women, but not many studies show how physical standards or perceptions of physical standards affect attitudes and experiences of servicemembers. While tests such as those conducted at IOC and RAP week represent a screening mechanism for those courses, the ACFT predicts fitness levels required to perform combat duties. Studies consistently support the design of the ACFT as a test that predicts combat soldier task performance in both men and women.[33] Female performance on the ACFT supports the conclusion made by several scholars who argue that overall, there are no specific physical requirements associated with combat that women have consistently failed to meet.[34]

Moreover, studies have found that certain differences in performance do not impede employment in physically demanding occupations.[35] Ran Yanovich et al. studied the Israeli Defense Forces to determine whether gender-integrated basic training can lessen gender differences in certain parameters of physical fitness relevant to mil-

---

[30] Agnes Gereben Schaefer et al., *Implications of Integrating Women into the Marine Corps Infantry* (Santa Monica, CA: Rand, 2015), 22; and *Women in the Military: Deployment in the Persian Gulf War* (Washington, DC: General Accounting Office, 1993).
[31] Ellen Haring and Megan Mackenzie, "Marine Corps Studies Miss the Mark," Women in International Security, 14 October 2015.
[32] Haring and Mackenzie, "Marine Corps Studies Miss the Mark."
[33] *Baseline Soldier Physical Readiness Requirements Study* (Iowa City: University of Iowa Virtual Soldier Research Program, 2020).
[34] Mackenzie, *Beyond the Band of Brothers*, 104.
[35] Delia Roberts et al., "Current Considerations Related to Physiological Differences Between the Sexes and Physical Employment Standards," *Applied Physiology, Nutrition, and Metabolism* 41, no. 6 (June 2016): 108, https://doi.org/10.1139/apnm-2015-0540.

itary task performance and identified strategies to reduce the gap in performance.[36] As such, though physical differences between men and women exist, women consistently surpass men in certain aspects of physical fitness and the gap in performance is not significant enough to invalidate participation in combat roles. Physical standards are therefore not reasons for excluding women but a justification for limiting their access to combat roles in the military.[37]

The Definition of Gender-Neutral Physical Standards

Little literature or research exists on the exact effect of gender-neutral physical standards on the perception of or attitudes toward women in combat arms, in part due to the absence of a concrete definition of the meaning of *gender neutral*. Because there is no set definition of "gender-neutral standards," critics argue that "all possible options for implementing 'gender-neutral standards' would have the effect of lowering requirements."[38] One definition of *gender-neutral standards* supported by DOD is that these standards "are based only on the physical capabilities required to perform the job, are the same for men and women, and should not differentially screen out a higher proportion of members of one gender who are, in fact, able to perform the job."[39] Interestingly, this definition may be impossible for women to meet for jobs that require physical capabilities that men, on average, are better suited for. Definitions of gender neutrality currently used by the military community therefore offer significant implications to policy outcomes.

Review of the existing literature on the definition of gender neutrality with regard to physical standards allows for the identification of two types of definitions or perceptions of gender neutrality. The first is largely dictated by the DOD and identifies gender-neutral physical standards as "tests and standards [that] are equally effective for both males and females."[40] In other words, gender is not a factor in decisions about the minimum qualifications for a job under this definition, and assessment is based on ability rather than gender. RAP exemplifies this type of gender-neutrality, as candidates are screened on standards that have not changed since the integration of women and do not distinguish between gender.

The second type of gender-neutrality refers to predictive bias in physical standards. Specifically, it perceives gender neutrality as subjective to the comparative ef-

---

[36] Ran Yanovich et al., "Differences in Physical Fitness of Male and Female Recruits in Gender-Integrated Army Basic Training," *Medicine and Science in Sports and Exercise* 40, no. 11 (November 2008): S654–59, https://doi.org/10.1249/MSS.0b013e3181893f30.
[37] Yanovich et al., "Differences in Physical Fitness of Male and Female Recruits in Gender-Integrated Army Basic Training."
[38] Jenna Grassbaugh, "The Opaque Glass Ceiling: How Will Gender Neutrality in Combat Affect Military Sexual Assault Prevalence, Prevention, and Prosecution?," *Ohio State Journal of Criminal Law* 11, no. 2 (Spring 2014): 339.
[39] Hardison, Hosek, and Bird, *A Review of Best Practice Methods*, xii–xiii.
[40] Szayna et al., *Considerations for Integrating Women into Closed Occupations in the U.S. Special Forces*, 203.

fort required for a woman or man to achieve the same results. By this definition, whether or not an activity is perceived as gender-neutral depends on "conceptions regarding gender, gender differences, and beliefs about the appropriateness of participation due to gender."[41] Performance tasks with a strong focus on anaerobic performance and strength may have the same standards for both men and women, but the standards for those tasks are easier for men to achieve with less effort than would be required of a woman. Predictive bias "refers to systematic error that occurs when a test/standard is a better predictor of performance of one group (e.g., men) than another group (e.g., women)."[42] Although the Army has moved away from the dual gender standards of the APFT and has adopted the gender-neutral ACFT, "gender-norming" is often treated as representing a dilution of physical standards by the military, while separate standards such as those of the APFT codify male physical fitness as superior, as men's minimum standards are significantly higher than women's.[43]

These two types of gender neutrality help construct the classification of gender-neutral standards as neutral in terms of outcomes or in terms of metrics. This classification is further discussed in the methodology section.

Argument

The independent variable in this study is the context in which gender-neutral standards are implemented, with the following independent variable indicators: 1) soldiers' perceptions of the effectiveness of a physical fitness standard in measuring combat-related fitness; 2) the degree of perceived gender-neutrality in the new physical fitness standard compared to the old one; and 3) the organizational culture of the unit or branch of Service to which a soldier belongs. The dependent variable is integration, which encompasses three indicators: cohesion, inclusion, and attitudes. All three are used to measure a level of integration qualitatively, while attrition rates before and after the implementation of the gender-neutral standard indicate the level of quantitative integration.

> Thesis: If soldiers perceive standards as gender neutral in terms of both metric and outcome, women in combat arms units experience higher levels of integration.

For the purposes of this study, *inclusion* is defined as equal participation and opportunity despite existing differences. Notably, *diversity* does not simply translate

---

[41] Nathalie Koivula, "Perceived Characteristics of Sports Categorized as Gender-Neutral, Feminine and Masculine," *Journal of Sport Behavior* 24, no. 4 (December 2001): 378.
[42] Szayna et al., *Considerations for Integrating Women into Closed Occupations in U.S. Special Operations Forces*, 206.
[43] Mackenzie, *Beyond the Band of Brothers*, 111.

to inclusion—the persistent aforementioned social myths surrounding women's participation in combat arms must be considered when considering the degree of existing inclusion.[44] *Cohesion* describes the strength of interpersonal bonds and the trust that often results from task-related competencies, including perceived or actual physical fitness measured by assessments. This definition encompasses the definition of military or unit cohesion with that of task and social cohesion, drawing on various scholars in the field of military psychology introduced in the literature review. Finally, this study defines *attitudes* as evaluations of a particular entity with some degree of favor or disfavor.[45] These evaluations, notably, can manifest themselves in behavior or stated opinion, meaning nonverbal responses and repeated behavior are just as important as explicit attitudes.

Analysis of the definitions of gender-neutrality and of the physical fitness test standards of the ACFT, RAP week, and IOC allows these standards as to be classified as gender-neutral in metric, outcome, or both. The metric-based definition of gender neutrality refers to a test like the ACFT, which applies the same metrics to measuring male and female performance. This definition fits under the DOD's general understanding of gender neutrality. The outcome-based definition stems from literature on performance-based gender neutrality and focuses on the outcome of the fitness test—the gap between male and female performance. Table 2.1 depicts how each examined test fits into the given definitions. The asterisk next to RAP week indicates the questionable categorization of this course as gender-neutral in both outcome and metric, as attrition rates and therefore outcomes vary between classes and years.

Table 2.1

Neutral metric Neutral outcome RAP week*	Neutral metric No Neutral outcome ACFT RAP Week IOC
No Neutral metric Neutral outcome APFT (normed Score)	No Neutral metric No Neutral outcome APFT (test results)

---

[44] Danielle R. Gamble, "Toward a Racially Inclusive Military," *Parameters* 50, no. 3 (Autumn 2020): 66.
[45] Alice H. Eagly and Shelly Chaiken, *The Psychology of Attitudes*, 1st ed. (San Diego, CA: Harcourt Brace Jovanovich, 1993).

This study expects to find higher levels of integration in cases where soldiers perceive standards as gender neutral both in terms of outcome and metric because women are measured and held to the same standard and are able to achieve the same outcomes as men. This study predicts that the organizational culture of a unit as well as the mission set of each unit might also contribute to these different perceptions. Specifically, scholars point out that group norms, a formal philosophy or mission, and espoused values form organizational culture.[46] As such, it is expected that the goals and philosophies—otherwise referred to as *espoused values*—of a unit define the context in which gender-neutral standards are implemented and thus the perception on the gender neutrality of a standard.

Concepts of perception and inclusion significantly contribute to the causal logic of this argument. Specifically, considering inclusion from the point of view of men and women may demonstrate a difference in perceptions of inclusion. Studies have found that "making women more aware of their stigmatized status led them to self-stereotype more in terms of their gender identity," highlighting their sense of belonging to a group of women, not necessarily to their unit.[47] For example, a study conducted by Cynthia L. Pickett and Marilynn B. Brewer found that when someone identifies highly with a group but also has the knowledge they are members of a "peripheral" of that group, they are more likely to engage in self-stereotyping and more negative stereotypical self-presentation.[48] Their research also concluded that "being near the edge of the ingroup . . . should be especially threatening and may result in . . . perceiving the outgroup as being more dissimilar to the ingroup than it really is."[49] As such, in combat arms branches or Services where women constitute a significant minority (5 percent or less of total population) and are thus a "peripheral," women's perception of the degree of gender neutrality of physical fitness requirements may differ from that of the men in the group.

Gender neutrality in terms of both outcome and metrics therefore has the possibility of dissolving the boundary that previously separated women from men and left women "near the edge of the ingroup." The challenge in female integration to combat arms and overcoming the male-oriented subculture fostered by the band of brothers myth thus lies in breaking down these barriers, where women may subtly take a marginal role.

The concept of perception is likely to impact the participants' view of inclusion and cohesion and shape their attitudes toward female integration in combat arms and must,

---

[46] Edgar H. Schein, *Organizational Culture and Leadership*, 4th ed. (San Francisco, CA: Jossey-Bass, 2010).
[47] Paul Hutchinson, Dominic Abrams, and Julie Christian, "The Social Psychology of Exclusion," in *Multidisciplinary Handbook of Social Exclusion Research*, ed. Dominic Abrams, Julie Christian, and David Gordon (Hoboken, NJ: John Wiley and Sons, 2007), 37.
[48] Cynthia L. Pickett and Marilynn B. Brewer, "The Role of Exclusion in Maintaining Ingroup Inclusion," in *The Social Psychology of Inclusion and Exclusion*, ed. Dominic Abrams, Michael Hogg, and Jose Marques (New York: Psychology Press, 2005), 102.
[49] Pickett and Brewer, "The Role of Exclusion in Maintaining Ingroup Inclusion," 105.

therefore, be considered. Jennifer Spindel and Robert Ralston, for example, found that "respondents with combat experience, who held/hold a higher rank, and who are currently serving are more likely to endorse a task-based conception of cohesion that ties cohesion to professionalism and competence, rather than social identity."[50] The methodology section describes ways in which perception is considered as a factor in the data collected from interviews.

Methodology

Through semistructured interviews of enlisted as well as commissioned members of the U.S. armed forces and retired servicemembers, the author gathered information on perceptions of women in the military under a gender-neutral physical standard. Where applicable, how that perception has changed since the introduction of a gender-neutral physical standard was analyzed. A snowball sampling technique was used to recruit participants to this study. The empirical section of this study is therefore based on qualitative data used to assess the effect of the independent variables on the dependent variables. Qualitative data analysis represents the best method to answer the research question due to the variety of ways in which respondents indicate perception and attitudes toward a given issue—as was previously mentioned, verbal and nonverbal indicators contribute to an understanding of a person's attitude on a subject. Moreover, qualitative data allows for a deeper understanding of factors that may contribute to these different attitudes such as experience, gender, and unit culture.

To conceptualize underlying patterns in interview responses, the author determined responses to each question to be either affirmative, neutral, or negative in relation to each dependent variable indicator. Interview responses were coded by analyzing verbal and nonverbal replies to the questions listed in, but not limited to, those given in the sidebar.

Specifically, the author searched for keywords that indicate a high or low level of inclusion or cohesion and a positive or negative attitude towards the integration of women. The endstate of each interview was to determine how cohesion fits into individuals' definitions of gender neutrality as well as to determine the level of inclusion women feel after the implementation or shift toward a gender-neutral standard, whether it be an actual or perceived shift.

The independent variable this study analyzed was the context in which physical standards were implemented. The change in a certain physical standard and the perception of that change served as the indicating variables for analyzing the social context. Perceptions of the effectiveness of physical fitness tests such as the ACFT, RAP week, IOC, and CFT/PFT in representing combat-related fitness also served

---

[50] Jennifer Spindel and Robert Ralston, "Taking Social Cohesion to Task: Perceptions of Transgender Military Inclusion and Concepts of Cohesion," *Journal of Global Security Studies* 5, no.1 (January 2020): 80, https://doi.org/10.1093/jogss/ogz045.

as indicators of the organizational culture and thus the context of implementation of gender-neutral fitness standards.

The dependent variable—level of integration—was determined by analyzing the dependent variable indicators of inclusion, attitudes, and cohesion. Notably, this study distinguishes between qualitative and quantitative integration, analyzing the level of dependent variable indicators qualitatively as well as looking at failure or attrition rates of women in each course or fitness assessment.

As discussed in the argument section, several factors can influence perceptions of women's performance and competence as reflected by gender-neutral physical standards. Prior experience in working with women and potential biases are very likely to affect these perceptions and will have to be considered when evaluating responses. Most importantly, male unit members' beliefs about the standards to which women are held will influence their perceptions of women's competence and therefore are likely to affect unit cohesion and therefore integration. Some U.S. military personnel believe that women are held to lower standards—this indication affects their perception of cohesion in a unit.[51]

This study is conducted under the assumption that cohesion does not cause effectiveness, as demonstrated by decades of research in group dynamics, organizational behavior, military sociology, sports psychology, and social psychology.[52] Although, generally, more-cohesive groups perform better than less-cohesive groups, researchers have yet to demonstrate a causal relationship between cohesion and unit effectiveness. However, cohesion is an indicator of the level of integration within a unit, which is why cohesion remains relevant to the study.

Case Study 1: Army
Demographics
Out of 14 Army servicemembers interviewed, 6 respondents were men and 8 were women, with 2 respondents serving as noncommissioned officers (enlisted). The respondents ranged in age, rank, experience, and operational unit, though all represented combat branches within the Army or were attached to a combat branch. Out of the eight women interviewed, three attempted RAP week.

Cohesion
Most interview responses from this data set indicated that cohesion was the dependent variable indicator with the least impact on the integration of women in combat arms. In other words, the context of implementation of a gender-neutral physical fitness test (independent variable) impacted cohesion the least. Specifically, cohesion

---

[51] Szayna et al., *Considerations for Integrating Women into Closed Occupations in the U.S. Special Forces*, 91.
[52] Robert MacCoun, "What Is Known about Unit Cohesion and Military Performance," in *Sexual Orientation and U.S. Military Personnel Policy: Options and Assessment* (Santa Monica, CA: Rand, 1993).

in Army combat arms units seemed to depend less on the perception of a certain physical fitness test and more on the overall competency of a unit member. Physical fitness, notably, contributes to competency, but was only mentioned as a prerequisite to unit cohesion. Several respondents shared the opinion that "as long as [the physical fitness of a female soldier] was good enough, then [cohesion] was fine."[53]

One respondent highlighted several factors that seem to have a significantly greater impact on unit cohesion, specifically trust in the organization, peers, and leadership.[54] Such cohesion is often developed through shared hardship in form of challenges that can only be accomplished if the unit works together as a team—"tough field problems, long field problems, [poor] conditions, events [soldiers] can't do by themselves."[55] Although physical fitness is certainly a component of such events, fitness standards are not.

Inclusion

The context in which gender-neutral physical fitness standards were implemented did not significantly affect quantitative inclusion of women in combat arms branches. The same cannot be said for qualitative inclusion, however, as the degree of perceived gender neutrality of a fitness test such as the ACFT continues to influence the way women are viewed by their peers in combat units.

As of March 2021, 62 women have graduated Ranger School.[56] With an estimated 35 percent female graduation rate, quantitative integration of women is comparable to the 45 percent graduation rate of men for the course as of 2019.[57] Although female Ranger School graduates still comprise a very small minority in the Army, the attrition rate is close to that of males attempting the course. Notably, a significant limitation to the data analysis in this study is the lack of access to attrition rates specific to RAP week, which is the physical fitness evaluation examined.

Women interviewed that have attempted or successfully completed RAP week expressed their perception of the physical fitness evaluations as being neutral in both outcome and metrics in relation to gender. One woman interviewed, for example, stated that although she felt some pressure as a female going through the course, "any man that goes through Ranger School thinks he is being targeted as well."[58] Moreover, leaders recognized women in Ranger School and at their operational units for their excellence at the same rate as their male peers. If a soldier met or exceeded the standard, no matter their gender, the Ranger instructors acknowledged

---

[53] 29 January Army officer interview.
[54] Male Army officer (anonymous), interview with author, 28 January 2021, hereafter 28 January Army officer interview.
[55] 28 January Army officer interview.
[56] Female Army officer (anonymous), interview with author, 10 March 2021, hereafter 10 March Army officer interview.
[57] "Airborne and Ranger Training Brigade, MCoE Fort Benning, GA, U.S. Army Ranger School" (PowerPoint presentation, Fort Benning, GA).
[58] 10 March Army officer interview.

this and often rewarded those who performed well.[59] The success rate of women at Ranger School and therefore passing RAP week, however, confirms RAP week as a neutral-outcome and neutral-metric physical fitness test with the population that attempts the course.

Attitudes

Given the demographics of the interviewed sample of respondents (the majority of whom were officers), many placed significant weight on the importance of leadership in determining the dependent variable indicator of attitudes. Several women noted that "everyone was really supportive" in a combat unit, which positively impacted their experience and therefore contributed to a welcoming environment.[60] Such responses indicate the context of implementation of gender-neutral physical fitness standards (independent variable) such as the ACFT. A male respondent further highlighted this sentiment by solidifying the condition that "if leadership takes an approach that physical fitness is important and gives everyone the opportunity to improve and be successful with it," leaders in units can alter the relative weight placed on the value of a gender-neutral or non-gender-neutral test.[61]

Both male and female respondents stressed the importance of diversity and embracing differences in the physical fitness capabilities of each sex. For example, several respondents pointed out that "there are other ways for women to show value than just physical fitness" and "we need to celebrate and embrace our differences."[62] Those who hold this opinion agree that meeting a male-oriented physical fitness standard or passing a physical fitness assessment such as RAP week is not a necessary condition for the integration and cohesion of women in combat arms. Interestingly, these responses counter the hypothesis that gender-neutral physical fitness tests lead to greater integration of women in combat arms. The respondents' backgrounds and experiences in specialized units may be a possible reason for this discrepancy. Another respondent highlighted the variety of missions conducted by any given unit, saying that a single standard makes sense for a single mission, but since there is a large variety of missions, fitness assessments should cater more to a mission-set than to a gender.[63]

These points were summarized well by Dr. Kyleanne Hunter, who stated that "calling something gender-neutral says we are going to make women fit the male

---

[59] 10 March Army officer interview.
[60] Female Army officer (anonymous), interview with author, 8 March 2021, hereafter 8 March Army officer interview.
[61] 29 January Army officer interview.
[62] Male Army officer (anonymous), interview with author, 1 February 2021, hereafter 1 February Army officer interview; and retired female Army officer (anonymous), interview with author, 20 February 2021, hereafter 20 February Army officer interview.
[63] 1 February Army officer interview.

standard and creates cognitive reinforcement that the 'man way' is the right way."[64] The findings described above thus support the creation of an alternative hypothesis that suggests gender-neutral physical fitness standards (either outcome- or metric-based) may not directly impede but certainly do not help cohesion or increase the quality of integration of women in combat arms units. In other words, gender-neutral physical fitness standards alone are not sufficient to increase integration and cohesion of women in combat arms roles.

Female respondents who graduated Ranger School shared the opinion that, at least initially, soldiers perceived women as incapable, believing women only received promotion or recognition due to politics.[65] As the exposure of soldiers to women in combat roles increased with time, their perception regarding women fulfilling these roles shifted to a more receptive and accepting attitude.[66] "The more touchpoints people have with seeing a demographic within a position," said one respondent, "the easier it is to cast aside a poor performer or two as an anomaly, not the standard."[67] In other words, as quantitative integration increases, qualitative integration can be expected to increase as well.

Overall, indicators of the dependent variable (integration) proved to be highest when all members of the team recognized that—regardless of a woman's performance on a physical fitness test—women are valuable additions to the team. Even women that failed to complete Ranger School and are currently serving in combat arms units without this qualification reported that they feel like valued members of a team. For example, one female respondent who did not graduate from Ranger School and was a platoon leader said that her fears of being judged or excluded from the team "were not validated" and that "it only matters that you can do your job and are a good leader."[68] Recognizing the inherent differences respondents may hold on the definitions of what it means to "pull your own weight" or "be a good leader," both Army and Marine Corps servicemembers share the opinion that the descriptions above imply being able to complete an assigned task without slowing the team down, requiring significant assistance or extra resources, or preventing the team from accomplishing the mission.

The timing of the change from a non-gender-neutral test (APFT) to a metric-gender-neutral test (ACFT) did not have a significant impact on the perceptions of the effectiveness of a given test. Accordingly, cohesion and integration were not significantly impacted by this policy change. As revisions to the ACFT continue, however, Army servicemembers may begin to shift their perception of the test if these revisions seem to be directed at areas in which women tend to perform worse than their male counterparts, such as the leg tuck. As of 23 March 2021, soldiers have the

---

[64] 20 February Army officer interview.
[65] 10 March Army officer interview.
[66] 10 March Army officer interview.
[67] 28 January Army officer interview.
[68] 8 March Army officer interview.

option to opt out of the leg tuck and perform a minimum two-minute plank instead.[69] Although interviews conducted in this study failed to adequately consider this change due to its recency, it is reasonable to infer that soldiers may view this change as catered to women, many of whom have struggled to perform a leg tuck.[70] This recent change warrants further study to determine the impact of the timing of changes to fitness tests on the perception of certain demographic groups.

In examining the responses relating to the independent variable indicator of organizational culture of the Army, interview responses demonstrated an accepting attitude toward embracing the differences described above. Specifically, most respondents who previously served or currently serve in the Army hold the opinion that "being able to pull your own weight without taking away from the team" has no effect on cohesion or the perception of women in combat arms.[71] In other words, respondents considered it sufficient for a woman in combat arms within the Army to be "good enough" or complete the task (physical or otherwise) to standard without "being a burden" to the team.[72] This view differs from that held by the Marines interviewed, which is discussed in further detail in the next section.

Case Study 2: Marine Corps
Demographics
The sample consists of three male Marines and three female Marines, all of whom had experience in combat units in the Marine Corps. Though none of the female Marines interviewed attempted the Marine IOC, inferences may be made from male experiences with women in The Basic School (TBS) and the IOC to determine the effect of the independent variable on dependent variable indicators examined.

Cohesion
Data analysis demonstrated that the context in which certain physical fitness standards are implemented does not significantly impact cohesion in the Marine Corps. Differences in the scales on which male and female Marines are evaluated do not necessarily matter to many Marines in combat roles. They recognize that "realistically, the things measured [on the fitness tests] are not the most important."[73] This attitude suggests that, contrary to this study's expectation, some Marines perceive their physical fitness tests as relatively irrelevant and/or ineffective at measuring combat effectiveness, which is an attitude shared by some in the Army toward the ACFT.

---

[69] "ACFT 3.0," Army.mil, 23 March 2021.
[70] Haley Britzky, "Lawmakers Are Mad About the Army's New Fitness Test and Think the Leg Tuck Is Useless," *Task & Purpose*, 21 October 2020.
[71] 28 January Army officer interview.
[72] 29 January Army officer interview.
[73] Male Marine officer, interview with author, 20 February 2021, hereafter 20 February male Marine officer interview.

A male interviewee claimed that men in the Marine Corps "just want to see a female operate on their level and they get excited to compete [with a woman] at that level," further noting that "if someone is physically at your level, you [as a male Marine] are more receptive and less critical to potential deficiencies they may have."[74] Another respondent, however, claimed that working with a woman who could not meet the male standard on a physical fitness test "did not have a big impact on how women were viewed," and as a superior there would be no difference "in how [he] would treat males that could not meet the standard."[75] In contrast to this male officer's opinion, female Marine officers expressed the view that they felt significant pressure to perform as well as or better than their male counterparts because "some people are very stuck in a certain mentality," though "generally people treated each other very professionally." For women as for men in the Marine Corps, exceptional physical fitness is "an entry level price to pay" to establish unit and task cohesion.[76]

Overall, then, men and women are expected to perform at the same level even if that expectation is not codified in Marine Corps regulations. Unlike Army servicemembers, Marines also do not think it enough for a woman to just "not be a liability."[77] On the contrary, respondents mentioned that women in the Marine Corps are expected to perform to the same standard or higher than the men in the unit, even if that is not necessary for their mission set.[78] Other than the organizational value that "officers need to be better than their Marines," no matter what their occupation, men tend to judge women in the Marine Corps based on their physical fitness.[79]

Inclusion

Quantitatively, women in the Marine Corps experience significantly less inclusion than men do. Data was unavailable regarding the female attrition rates at IOC, but research indicates that out of 10 women to attempt the IOC since 2016 only 2 (20 percent) have graduated.[80] In contrast, approximately 70 percent of men who attempt the IOC graduate the course.[81] As such, the gender-neutral (in terms of metric) standards of the IOC result in less quantitative integration of women as illustrated by the high attrition rate for women attempting the course (80 percent). Notably, out of the interviewed population, both men and women in the Marine Corps perceive the

---

[74] 20 February male Marine officer interview.
[75] Male Marine officer, interview with author, 19 March 2021, hereafter 19 March male Marine officer interview.
[76] Female Marine officer, interview with author, 16 April 2021, hereafter 16 April female Marine officer interview.
[77] Retired male Marine officer, interview with author, 12 February 2021.
[78] Female Marine officer, interview with author, 20 February 2021, hereafter 20 February female Marine officer interview.
[79] Retired female Marine officer, interview with author, 12 February 2021, hereafter 12 February retired female Marine interview.
[80] Philip Athey, "First Female Infantry Marine Officer Leaves Corps as Commandant Calls for More Women at Infantry Officer Course," *Marine Corps Times*, 25 February 2020.
[81] "Infantry Course Completion Rates," *Military Times*, accessed 16 April 2021.

IOC as gender neutral in terms of metrics because "men and women are held to the same standard."[82]

Certain jobs within the Marine Corps actually place women at an advantage, but even in these positions female Marines are expected to be better, faster, and stronger than their subordinates and many of their peers. When operating in the human intelligence field, for example, gender poses an advantage because "they need[ed] women to do certain missions and fulfill certain roles that men were incapable of doing."[83] Moreover, women in such units note that their credibility was "based off intelligence, and not [your] gender or physical test score," though "physical fitness is definitely very important."[84]

In addition to interviews conducted throughout the research process, several other examples support the argument that attitudes of leaders in the Marine Corps determine the context in which the force implements gender-neutral physical standards. General Robert H. Barrow, for example, testified before Congress and argued that female integration itself would "destroy the Marine Corps . . . something no enemy has been able to do in over 200 years."[85] A video of his testimony was said to be largely circulated throughout the Marine Corps, undoubtedly influencing the opinion of many Marines. Although this testimony concerned the overall integration of women into combat arms branches, not fitness standards, it contributed to the more scrupulous attitude toward women in the Corps and a belief that women had to earn their spots as Marines.

As time passed and women became a part of the Marine infantry or fulfilled other combat roles, however, there have been apparent changes in attitudes of leaders and subordinates alike. As one respondent noted, "Overall, the Marine Corps will always be responsive [to change]—Marines just grumble sometimes."[86] Another respondent noted that "men that go through the course are pretty shocked by female capability," meaning women have the opportunity to prove themselves and potentially alter the perception of those around them.[87]

Where a gender-neutral standard does not exist, Marines tend to compare both men and women on the male scale. In other words, everyone is informally compared to one another on the same scale (the standard performance of men in the unit) even if they are not being evaluated on the same scale. This is the case for the Combat Fitness Test (CFT) and Physical Fitness Test (PFT)—the two fitness tests in the Marine Corps.[88] This conclusion is further supported by one respondent's opinion that "a

---

[82] 20 February male Marine officer interview; and 20 February female Marine officer interview.
[83] Elizabeth Chang, interview with Lauren Finch Serrano, in "The Few, the Proud: Women Marines Define Themselves," *Washington Post*, 2 November 2017.
[84] 16 April female Marine officer interview.
[85] Paul Szoldra, "General in Charge of Female Recruit Training Once Suggested Gender Integration Would Destroy the Marine Corps," *Task & Purpose*, 1 May 2020.
[86] 16 April female Marine officer interview.
[87] 20 February female Marine officer interview.
[88] 20 February female Marine officer interview.

universal metric does not change the culture of the unit, just removes the argument against women."[89] Therefore, from conversations with Marines, it is inferred that the most significant barrier to integration of women into the combat environment is mental and that physical standards or abilities may not necessarily sway the opinion of many Marines regarding women in combat occupations.

Summary and Discussion

Whereas Army servicemembers agreed that physical standards and performance were only as relevant as the requirements of the specific occupation or job a woman belongs to, women and men in the Marine Corps shared the opinion that even if a task or certain level of fitness was not necessary for a job, both women and men were expected to achieve that higher level of fitness.

While some women in combat arms in the Army feel less pressure or do not feel judged even when failing to complete Ranger School, women in the Marines—notably, even those not in combat occupations—feel they have "a lot to prove in the Marine Corps" and fitness is "one of those areas that a lot of male Marines tend to judge women on."[90] The perception of the importance of physical fitness performance of women and men in noncombat occupations also differs from the perception held by those in the Army. The degree or type of gender neutrality of a test does not seem to matter, as Marines subconsciously evaluate both sexes on the same scale.

Marines also expressed the sentiment that physical fitness is taken more seriously in the Marine Corps as part of the culture. Specifically, one Marine noted the significant difference in the size of the Marine Corps when compared to the Army, highlighting the importance of physical fitness test scores in promotion boards and evaluations due to the extremely limited number of high-ranking command positions in the Corps.[91] For example, this Marine noted that a score of 250 points or more is required of any officer before attending officer-specific training, though the passing score on the test is 150.[92] Similarly, to be promoted or at least be competitive for O-5 command, an officer must score in the 270–85 range.[93] That is because these tests are some of the very few things that can distinguish one officer from another in an extremely selective promotion process where very few officers have the opportunity to serve in a given capacity. Although fitness test scores are a useful metric for evaluating the readiness of officers, they seem to hold too much weight in individuals' perceptions of women and their capabilities.

Moreover, officers experience greater pressure to perform exceptionally on physical fitness standards than enlisted Marines do, which Marines recognize as "a

---

[89] 12 February retired female Marine interview.
[90] 20 February female Marine officer interview.
[91] 19 March male Marine officer interview.
[92] *Marine Corps Order 6100.13A, Physical Fitness and Combat Fitness Tests* (Washington, DC: Headquarters Marine Corps, 13 March 2019).
[93] 19 March male Marine officer interview.

cultural value of the Corps."[94] Even the highest-ranking Marines are aware of this culture and continue to promote it. Retired general James N. Mattis noted that officers need to "be as physically tough as [their] toughest troops," saying "don't try to 'wow' [soldiers] with your knowledge if you can't keep up with them on any run or outrun them."[95] These comments support the argument that exceptional physical fitness is the baseline expectation in the Marine Corps without which women cannot experience full integration into combat arms.

The Army is a larger force with different requirements of its servicemembers, and the Marine Corps is generally much more focused on specific tasks related to physical fitness.[96] Moreover, the number of women seeking to participate in combat occupations in the Marine Corps is extremely low, supporting policymakers' concerns that there are "too few . . . stalwart young ladies . . . charging into this [combat occupations]."[97] A Marine working in the Infantry Training Battalion noted that out of the women participating in this basic training required of enlisted Marines in the infantry, none expressed interest in joining the infantry after going through the training course.[98] This low interest can be due to a number of factors, such as the organizational culture of the Marine Corps and recruiting procedures. Notably, this highlights that the dependent variable indicator of quantitative integration is partially affected by the low interest of women to serve in combat arms branches in the Marine Corps.

Conclusion and Recommendations

Data gathered from interviews with Marine and Army servicemembers supports this study's hypothesis, but not to the extent expected. The hypothesis stated that if standards are perceived as gender neutral in terms of both metrics and outcome, women in combat arms units will experience significantly higher levels of integration. Further study refined this hypothesis by indicating that gender-neutral physical fitness standards alone are not sufficient for qualitative integration of women in combat arms. Metric-based gender-neutral physical standards may help create more favorable attitudes and contribute to cohesion in the unit as long as there is support from leadership and internal gendered scales are not introduced for the purposes of recognition or promotion. Policymakers need to consider the following factors to ensure gender-neutral fitness tests have the desired effect: organizational culture of the unit, context of implementation of a standard, and perceptions of gender neutrality.

There are several areas that future studies on the impact of gender-neutral physical fitness standards on the integration of women in combat arms should focus. First,

---

[94] 20 February male Marine officer interview.
[95] David Choi, "Former Defense Secretary Jim Mattis Offers Advice for Young Leaders: 'Know Your Troops'," *Business Insider*, 1 February 2021.
[96] 19 March male Marine officer interview.
[97] Mattis remarks.
[98] The women in this course were all enlisted and were part of a pilot program designed to test and evaluate the integration of women into the Marine Corps Infantry. 19 March male Marine officer interview.

this study could be improved by expanding the representative sample to consider age and rank of respondents. Unfortunately, data collection did not involve conversations with enlisted women—a group that policymakers such as the former secretary of defense, Mark Esper, identified as a crucial stakeholder. Analysis of perceptions of lower enlisted servicemembers lends to comparison with the perceptions of officers in the Service and could be used to further study the impact of background experiences and organizational culture on attitudes toward gender-neutral fitness tests or the integration of women.

Second, this study could be applied to Special Operations Forces (SOF) to compare the integration of women in these units with integration of women in conventional units. Responses from participants in the data set suggested that when serving with or in a SOF unit, women felt greater belonging and qualitative inclusion. Keeping in mind the goal set out by policymakers to make "the best possible use of talents and capabilities of women who sought combat positions," it is clear that DOD policies require alteration to support this goal and embrace differences in physical strengths of each gender.[99]

As one of the first scholarly attempts to establish a relationship between the perception of physical fitness standards and the integration of women in combat arms, this chapter is bound by limitations. More importantly, however, this study prompts further research in various spheres relating to such a pertinent issue. Additional research in this field will help the U.S. military move toward a more effective talent assessment that gives equal opportunity to all genders and facilitates integration in the best manner possible.

---

[99] Carter, "No Exceptions."

Starting Interview Questions

These are the questions the author asked of anonymous servicemembers in interviews conducted as part of the research for this paper.

1. Describe your occupation. Do physical fitness standards have a significant impact on your occupation?
2. In your opinion, do gender-neutral physical standards positively impact the integration of women in combat arms units?
3. Do you perceive the current ACFT/Army Ranger School/Marine Corps Infantry Officer Course physical standards as gender neutral?
4. How would you define *gender neutral*?
5. How have your experience and interactions with women (if applicable) differed at various duty stations and time periods?
6. (*For women that successfully completed the Army Ranger School or Marine Corps Infantry Officer Course*) Were you treated differently before and/or after graduating from [respective school/course]? Did you notice any significant changes in the cohesion of the unit you belonged to?
7. Was there a shift in the attitude and/or effectiveness of your team or unit after the implementation of the ACFT/after 2015?
8. In your experience, has a gender-neutral fitness assessment improved or degraded unit cohesion and the integration of women in combat arms units?
9. (*For graduates of Army Ranger School or Marine Corps Officer Infantry Course since 2015*) Did you notice a difference in the way women were perceived or treated during and after completion of the course/school?
10. In your opinion, has the perception of women changed since the implementation of the ACFT/removal of the Combat Endurance Test (or at the Marine Corps Officer Infantry Course)?
11. How well did your unit accomplish goals together (*task cohesion*)? Did you feel like part of the team (*unit cohesion*)?

PART 3

Gender and Violence

Breaking a Vicious Cycle
Systemic Endorsement of Violence
Against Women in El Salvador
by Captain Elizabeth Jane Garza-Guidara, USAF*

Introduction

From 2007 to 2012, El Salvador had the highest reported femicide rate in the world and currently has the highest femicide rate in Latin America.[1] Despite these alarming statistics, gender-based violence against women is not regarded as a serious problem by many people in El Salvador. Post–civil war (1992 to present) conditions sowed seeds for the normalization of violence in El Salvador, especially violence against women. The postwar environment also exacerbated the dehumanization and subjugation of women in El Salvador, which has limited their access to justice and undermined the credibility of their legal testimony against their abusers.

By investigating how women's legal and structural inequalities are compounded by the normalization of violence as a key component of gender relations, this paper seeks to answer the following question: What has contributed to Salvadoran law enforcement and judicial systems' legitimization and endorsement of violence against women in El Salvador? Answering this question will help forge a comprehensive understanding of how the embrace of gender-based violence as socially acceptable reinforces the impunity of judicial and law enforcement institutions. Furthermore, the objective of this paper is twofold: 1) to investigate how barriers to justice for women have normalized gender-based violence as a part of everyday life and 2) to examine how the normalization of gender-based violence compels government officials to discredit and disregard women's testimony in court. Exploring this phenomenon also sheds light on how systemic discriminatory treatment of female victims not only constitutes a violation of women's rights, but is also noncompliant with El Salvador's WPS national action plan (NAP).[2]

Postwar El Salvador
and the Normalization of Gender-Based Violence

Postwar El Salvador created fertile ground for the normalization of gender-based violence. Radical levels of violence during the war impacted Salvadorans in two ways:

---

* The views expressed in this chapter are solely those of the author. They do not necessarily reflect the opinion of Marine Corps University, the U.S. Marine Corps, the U.S. Navy, the U.S. Army, U.S. Army War College, the U.S. Air Force, or the U.S. government.
[1] United Nations, "Deputy Secretary-General Applauds El Salvador for Implementing Spotlight Initiative, Tackling Highest Rate of Femicide in Latin America," press release, 8 December 2020.
[2] *National Action Plan for Resolution 1325: "Women, Peace and Security,"* 2017–2022 (San Salvador, El Salvador: Ministry of Foreign Affairs, 2017), unofficial translation, funded by ARC DP160100212 (CI Shepherd), hereafter El Salvador NAP.

1) it increased their tolerance for violence and 2) they embraced it as a normal means of problem solving.[3] The deep integration of violence into Salvadoran society impairs individuals' abilities to recognize certain acts as violent.[4] This has also contributed to the emergence of socially constructed levels of "acceptable" violence, which includes beating spouses, partners, or children with belts or pieces of wood.[5]

In El Salvador, violence is viewed as a natural form of male behavior toward women and as a critical element of gender relations.[6] Furthermore, this gender norm not only naturalizes the abuse of women but also strengthens biases that prevent women from obtaining justice. In this context, *machismo* (which is based upon an exclusionary and highly masculinist logic) is a vehicle for informing and reproducing gendered violence.[7] Multisided violence (a combination of structural, symbolic, political, and gender violence) prevents formal institutions from advocating for women's equality and reduces institutional responsiveness to women's rights.[8] This can be attributed to the fact that multisided violence further legitimizes violence against women as an integral part of men's interaction with women. Hence, gender norms and multisided violence have prompted members of the Salvadoran judiciary and police force to routinely react to violence against women with impunity. Ultimately, the minority of women who report violence are faced with either the unwillingness or inability of Salvadoran judicial authorities to provide justice.[9]

In addition to the ramifications of gender norms and multisided violence, women face yet another obstacle to obtaining justice against their abusers. The privacy of family life in El Salvador is used to legitimize limited intervention of the state in domestic violence cases, which continues to subjugate women as second-class citizens.[10] Because of this, police often encourage women to forgive and reunite with their male abusers instead of holding them legally accountable. According to a female Salvadoran domestic violence victim, she was "standing in front of the police, bleeding, and the police said, 'Well, he's your husband'."[11] Another Salvadoran woman stat-

---

[3] Karen Musalo, "El Salvador—A Peace Worse than War: Violence, Gender and a Failed Legal Response," *Yale Journal of Law and Feminism* 30, no. 1 (2018): 18.
[4] Mo Hume, " 'It's as if You Don't Know, Because You Don't Do Anything About It': Gender and Violence in El Salvador," *Environment & Urbanization* 16, no. 2 (October 2004): 64, https://doi.org/10.1177/095624780401600223.
[5] Hume, " 'It's as if You Don't Know, Because You Don't Do Anything About It'," 64.
[6] Cecilia Menjívar and Shannon D. Walsh, "Impunity and Multisided Violence in the Lives of Latin American Women: El Salvador in Comparative Perspective," *Current Sociology* 64, no. 4 (July 2016): 586–602, https://doi.org/10.1177/0011392116640474.
[7] Hume, "'It's as if You Don't Know, Because You Don't Do Anything About It'," 64.
[8] Menjívar and Walsh, "Impunity and Multisided Violence in the Lives of Latin American Women," 594.
[9] Rachel Dotson and Lisa Frydman, *Neither Security nor Justice: Sexual and Gender-based Violence and Gang Violence in El Salvador, Honduras, and Guatemala* (Washington, DC: Kids in Need of Defense, 2018), 3.
[10] Mo Hume, "The Myths of Violence: Gender, Conflict, and Community in El Salvador," *Latin American Perspectives* 35, no. 5 (September 2008): 66, https://doi.org/10.1177/0094582X08321957.
[11] *Women on the Run: First-Hand Accounts of Refugees Fleeing El Salvador, Guatemala, Honduras and Mexico* (New York: United Nations High Commissioner for Refugess, 2015).

ed that when the police came to her home after she called them to report domestic violence, they said that she and her husband could resolve it between themselves.[12]

When victims of gender-based violence reside in gang-controlled areas, crimes have an even higher probability of resulting in impunity.[13] Women living in gang-controlled areas throughout El Salvador are vulnerable to being forced into sexual relationships with gang members. If a woman refuses to comply with this demand, the gang member threatens to sexually assault her or to injure or kill her family members. Police, prosecutors, and judges in El Salvador often assume that females targeted by gangs (or from gang-controlled areas) have ties with gangs. This bias is then utilized to rationalize discrimination against these women, reject their reports, or "publicly question the credibility of their accounts."[14]

The widespread acceptance of this bias is clearly illustrated when state officials and the media commonly criminalize female homicide victims from poor areas by declaring that they were involved with gangs. This leads to "a gender-motivated murder being classified as gang conflict rather than femicide."[15] Furthermore, it can be hypothesized that the subjugation of women and the normalization of gender-based violence in El Salvador fuels systemic discriminatory treatment of women by the judiciary and police.

El Salvador's Abortion Law: An Expression of Gender-Based Violence

Domestic violence is not the only kind of violence that women are routinely subjected to in El Salvador. The criminalization of abortion in El Salvador exacerbates gender-based violence in two ways: 1) the discriminatory and violent treatment of "suspects" and 2) the violation of women's legal, human, and reproductive rights. Prior to the civil war peace accords, El Salvador permitted legal abortions in three circumstances: when the life of the mother was at risk; when the pregnancy was the result of rape; or when the fetus "had deformities incompatible with extrauterine life." In 1998, a new law criminalized abortion under all circumstances and extended the criminal sentence.[16]

As a result, Salvadoran women cannot legally terminate their pregnancies if their lives are at risk and/or if they are facing pregnancy-related complications.[17] This not only violates women's rights to health, but also means that a majority of women who

---

[12] *Women on the Run*, 26.
[13] Dotson and Frydman, *Neither Security nor Justice*, 3.
[14] Dotson and Frydman, *Neither Security nor Justice*, 9.
[15] Dotson and Frydman, *Neither Security nor Justice*, 10.
[16] Jocelyn Viterna and Jose Santos Guardado Bautista, "Pregnancy and the 40-Year Prison Sentence: How 'Abortion Is Murder' Became Institutionalized in the Salvadoran Judicial System," *Health and Human Rights* 19, no. 1 (June 2017): 84, https://www.jstor.org/stable/10.2307/90007917.
[17] "From Hospital to Jail: The Impact on Women of El Salvador's Total Criminalization of Abortion," *Reproductive Health Matters* 22, no. 4 (November 2014): 52, https://doi.org/10.1016/S0968-8080(14)44797-9.

are raped are forced to carry their pregnancies to term.[18] Over the past 20 years, approximately 181 women who experienced obstetric emergencies were prosecuted for abortion or aggravated homicide.[19] This reflects a deeply rooted *machista* attitude that embraces the following rationale: because a woman's role is only of a mother, the fetus is given priority over the health and life of the pregnant woman. In El Salvador, women "are considered to be morally incapable" of making a decision to save the baby or save herself, which "denies them their status as human beings and violates their right to life as granted in the Constitution of El Salvador, Article 2."[20]

During the initial wave of antiabortion activism, "abortion" and "homicide" were often used interchangeably in El Salvador. Hence, this distortion between the distinct differences between homicide and abortion have become institutionalized in the Salvadoran judicial system.[21] An embodiment of this development entailed systematic gender discrimination where the "state aggressively pursues the woman's prosecution instead of the truth at every stage of the judicial process."[22]

In cases where women were found guilty of having an abortion, there have been indications of dire violations of the right to due process (specifically the right to the presumption of innocence).[23] If a woman went to a public hospital because she was bleeding, she lost her right to be assumed innocent and was investigated. This escalated into an assumption that every woman who is bleeding has had an abortion and then must be investigated and prosecuted.[24] The criminalization of abortions under all circumstances also makes women even more vulnerable to a higher level of violence. Handcuffing women suspected of having abortions to their hospital beds is common and is a form of violence against women.[25]

Hospitals are not the only settings where women are dehumanized in El Salvador. Courtrooms also fulfill this purpose. Judicial officials who prosecuted women for abortions justified their decisions by declaring that the defendants "violated social expectations of motherhood." For example, they argued that mothers should always know when they are pregnant and should seek to protect their unborn babies even when suffering a dire medical crisis and losing consciousness. Judges also often only admitted evidence that supported the guilt of the defendant and "systematically excluded evidence that supported the women's testimony."[26] Police who investigated alleged abortions were often notorious for gathering evidence that would incriminate women. In some cases, evidence was obtained illegally.

---

[18] *Marginalized, Persecuted, and Imprisoned: The Effects of El Salvador's Total Criminalization of Abortion* (New York: Center for Reproductive Rights, 2014).
[19] Joe Parkin Daniels, "El Salvador Abortion Laws on Trial in Case of Woman Jailed after Miscarriage," *Guardian*, 12 March 2021.
[20] "From Hospital to Jail."
[21] Viterna and Bautista, "Pregnancy and the 40-Year Prison Sentence," 90.
[22] Viterna and Bautista, "Pregnancy and the 40-Year Prison Sentence," 88.
[23] *Marginalized, Persecuted, and Imprisoned*, 51.
[24] "From Hospital to Jail," 57.
[25] *Marginalized, Persecuted, and Imprisoned*, 56.
[26] Viterna and Bautista, "Pregnancy and the 40-Year Prison Sentence," 89.

In 20 abortion cases examined by Harvard Sociology Professor Jocelyn Viterna in 2017, the Salvadoran authorities routinely ignored evidence in order to pursue guilty verdicts.[27] A presumption of guilt was evident in all 20 cases. Police and government lawyers failed to recognize the possibility of an unanticipated birth and instead argued that defendants threw their newborns into the latrine. Despite the fact that forensic specialists' autopsies revealed fetal abnormalities that were likely to have caused natural stillbirths, they utilized flawed logic and unreliable tests to conclude that the babies were killed. Salvadoran judges not only accepted inconsistencies in forensic reports but also excluded information that would have supported the women's testimony.[28]

Moreover, women who have miscarriages are extremely vulnerable to being prosecuted for illegal abortions (and can be later charged guilty of aggravated homicide) and can be sentenced to serve 30 to 50 years in prison.[29] This unfortunate situation occurred during the *Manuela v El Salvador* case. On 10 March 2021, the Inter-American Court of Human Rights heard arguments in *Manuela v El Salvador*, which is a "landmark case that could shape policy and debate on abortion across Latin America." This was the first time that an international court challenged El Salvador's draconian approach to abortion.[30]

Starting in 2006, Manuela regularly sought medical care for an unknown chronic medical condition. On 26 February 2008, Manuela felt an intense abdominal pain, went to the outdoor latrine (where she felt like something exited her), and fainted. Her parents took her to the hospital. That same day, the hospital sent a report to the Public Prosecutor's Office accusing Manuela of abortion. The next day, police officers interrogated Manuela and conducted a search of her family's home. Police officers asked Manuela's illiterate father (Juan) to sign a document but did not explain to Juan that he had signed a criminal complaint against his own daughter. Juan's complaint and the complaint filed by Manuela's physician from the hospital were then used as evidence in the trial against Manuela.[31]

At the time of Manuela's arrest, authorities failed to fulfill minimal procedural guarantees. Manuela was not given sufficient time to prepare her defense and could not freely and privately communicate with her attorneys. Despite heinous violation of Manuela's legal rights, she was convicted and sentenced to 30 years in prison for aggravated homicide.[32] Manuela eventually died of cancer in prison without having proper treatment for her condition.

---

[27] Jocelyn Viterna, "The Real Reason El Salvador Jails Women for Stillbirths?: It's Called 'Moral Panic'," *Los Angeles Times*, 30 July 2017.
[28] Viterna, "The Real Reason El Salvador Jails Women for Stillbirths?"
[29] "From Hospital to Jail," 52.
[30] Daniels, "El Salvador Abortion Laws on Trial in Case of Woman Jailed after Miscarriage."
[31] *Marginalized, Persecuted, and Imprisoned*, 37.
[32] *Marginalized, Persecuted, and Imprisoned*, 38.

The Dire Impact of El Salvador's Abortion Law on Women and Girls

The *Manuela v El Salvador* case reveals that the women accused of abortion are among the poorest women in the nation.[33] To make matters worse, the criminalization and repression of poor women and girls has become commonplace, to include cases of nine-year olds becoming mothers after rape by a family member. While perpetrators of sexual and physical violence are not convicted for their crimes, women are punished for miscarrying.[34] In addition to a clear violation of women's civil rights, the abortion ban threatens women's well-being in El Salvador. Girls as young as nine have been denied therapeutic abortions and three out of every eight maternal deaths in El Salvador are pregnant adolescents who commit suicide.[35]

Gender-Based Violence and Women, Peace, and Security

In order to formulate solutions geared toward protecting women and girls from El Salvador's draconian abortion law and gender-based violence, it is necessary to examine these phenomena through the lens of the WPS program. UNSCR 1325 established WPS in 2000 to champion the criticality of the integration of a gender perspective for peace building and effective military and police operations.[36] In 2017, El Salvador adopted its WPS NAP.[37] To evaluate the extent to which El Salvador is fulfilling its WPS NAP, it is imperative to investigate two phenomena: 1) the degree to which gender equality principles have been integrated into El Salvador's laws and regulations and 2) the extent to which there is political will and commitment to integrate WPS principles into the judiciary and police force.

According to El Salvador's WPS NAP, the Salvadoran state declares its commitment to promoting and supporting initiatives to protect women and girls' human rights. El Salvador's Five-Year Development Plan (2014–19) is nested within the WPS NAP. The Five-Year Development Plan seeks to incorporate a gender approach in the following domains: education, citizen security, access to health, human rights, the maintenance of the historical memory and construction of peace, and progress toward a state aligned with civil society and serving as a promoter of human rights.[38]

On 25 November 2010, the Comprehensive Special Law for a Life Free of Violence for Women was approved. The objective of this law is twofold: 1) to regulate

---

[33] Michelle Oberman, "The Consequences of El Salvador's Abortion Ban: A Cautionary Tale of How Criminalization Impacts Women Across Class Lines," *Guernica*, 11 January 2018.
[34] Belén Fernández, "The Country Where Having a Miscarriage Can Land You in Prison," *Jacobin*, 31 January 2020.
[35] Lisa Kowalchuk, "The Unspeakable Cruelty of El Salvador's Abortion Laws," *The Conversation*, 11 April 2018.
[36] Chantal de Jonge Oudraat et al., *Enhancing Security: Women's Participation in the Security Forces in Latin America and the Caribbean* (Washington, DC: Women in International Security, 2020).
[37] El Salvador NAP.
[38] *National Action Plan for Resolution 1325*, 15.

the detection, prevention, and punishment of violence against women and 2) to ensure the care, protection, and reparation for the rights of victims. The law defines the following as crimes: femicide, obstruction to justice, and other "expressions of violence against women." This law also created the National System of Statistical Data and Information on Violence Against Women.[39] However, this law fails to achieve two things: 1) to recognize a lack of judicial and law enforcement impartiality as obstruction to justice and 2) to recognize the treatment of abortion suspects (such as being handcuffed to hospital beds) as expressions of violence against women.

Another law seeking to protect women is the Equality, Equity, and Eradication of Discrimination Against Women, which was established in March 2011. This law seeks to guarantee women the full exercise of citizenship and human rights, which include the following: health, social, and cultural rights; education; employment opportunities; and political participation. This law also requires the creation of an entity responsible for upholding women's rights under the Office of the Attorney General of the Republic.[40]

However, El Salvador's legal and political rhetoric on WPS does not align with its implementation of WPS. El Salvador's judiciary and police force minimally employ gender perspectives. Consequently, despite the creation of a WPS NAP, a Five-Year Development Plan, and laws geared toward upholding women's rights, the tenets of these strategic policies and laws are often disregarded by judges and police officers. Instead of exercising rights as free and equal citizens that are clearly stated in the aforementioned laws and the WPS NAP, women are caught in a vicious cycle of subjugation.

Policy Recommendations

In order to tackle the root causes of the systemic legitimization and endorsement of violence against women in El Salvador, it is critical to employ a holistic approach that addresses the weaponization of traditional gender roles to maintain women's subordinate status in society. With this in mind, implementing eight policy recommendations at the national and at the municipal and neighborhood levels throughout El Salvador will contribute to this effort. The actions outlined below need to be taken at the national level.

1. Ensure that all participating institutions outlined in the WPS NAP integrate the NAP into their strategic plans, including the Ministry of Justice and Public Security (which the judiciary and police force are a part of). Disseminate information related to the WPS NAP and make it available to civil society.
2. In addition to integrating the WPS NAP into strategic plans, all government agencies (including the judiciary and police force) should develop and execute their own unique WPS implementation plans. These WPS implementa-

---

[39] *National Action Plan for Resolution 1325*, 16.
[40] *National Action Plan for Resolution 1325*, 16.

tion plans should outline clear goals and milestones for measuring progress. In order to execute these plans, the National Civil Police, the Legislative Assembly, the Ministry of Justice and Public Security, and the Supreme Court should appoint Gender Advisors (GENADs) in order to achieve the following: aid with the development and implementation of organizational gender mainstreaming strategies; ensure that police operations have integrated a gender analysis; and advise on WPS education and training of police officers and judges.[41] Gender Focal Points (GFPs) should be appointed at lower echelons of these government agencies to ensure the implementation of WPS from top to bottom.

3. The Supreme Court of Justice needs to create and disseminate standards and protocol for guaranteeing the right to due process, avoiding prejudices, and respecting procedural guarantees.[42] For this to be effective, it is critical to create a comprehensive feedback loop that holds judicial officials accountable for failing to adhere to these standards and protocol. In order to achieve this, protocol must require the presence of a GFP during a court proceeding or a police operation involving a gender-based violence victim or a female abortion defendant. GFPs should be well educated on these standards and protocol (as well as national laws protecting women's rights) in order to notice and report violations to the Supreme Court of Justice GENAD. The Supreme Court of Justice GENAD will be responsible for coordinating necessary actions for violations and monitoring their status from start to finish (to include notifying judges of punishments for their actions). To further deter judges from violating protocol and standards, the Supreme Court of Justice should compile and disseminate reports of judges found guilty of protocol violations on a quarterly basis (to include listing punishments) to all members of the judiciary.

4. Institutionalize WPS training and oversight programs in the judicial and law enforcement sectors. The Salvadoran government needs to institutionalize WPS training and oversight for police, prosecutors, and judges to ensure sensitivity and application of due process during gender-based violence cases. Police personnel should also be routinely trained on the following: incorporating gender perspectives during operations; the guarantee of human rights (especially sexual and reproductive rights); addressing gender norms and stereotypes within judicial and law enforcement organizations in order to refrain from treating individuals based on these viewpoints; community policing strategies; and tactics that use violence as a last resort.[43]

5. Allocate required resources (personnel and financial) to ensure effective im-

---

[41] Oudraat, Haring, Islas, and Velasco, *Enhancing Security*, vi.
[42] *Marginalized, Persecuted, and Imprisoned*, 67.
[43] Dotson and Frydman, *Neither Security nor Justice*, 13–14.

plementation of El Salvador's WPS NAP. The Legislative Assembly must become a pivotal driver in implementing the country's WPS NAP by presenting routine NAP updates and passing budgets with dedicated funding for the NAP.[44]

6. Equip El Salvador's National System of Statistical Data and Information on Violence Against Women with a mechanism for neighborhood and municipal entities to directly input gender-based violence data into it. This mechanism should be used uniformly across government agencies. This will not only provide more accurate and current information about the degree and types of violence, but will also guide policy decisions, and monitor progress.[45]

7. Establish systematic monitoring and evaluation mechanisms for El Salvador's WPS NAP. Because GENADS are a critical component of WPS implementation oversight, GENADS should utilize the Latin America and Caribbean WPS assessment tool in order to measure progress in a concrete, data-driven manner. This tool measures progress on three levels via qualitative and quantitative indicators: political will; institutional policy and practice; and monitoring and evaluation.[46] After gathering and compiling data, GENADS should propose and execute corrective action plans that align with their organizations' respective WPS implementation plans. All of this information should be published in a report that is then disseminated to Ministry of Justice and Public Security, National Civil Police, Supreme Court, and Legislative Assembly leadership. Reports should be created and released on at least an annual basis.

At the municipal and neighborhood levels, WPS programs need to be implemented holistically and tailored to the specific needs of various communities. Furthermore, carrying out the policy recommendation below is critical to achieving this.

8. Implement thorough WPS public education programs that explore the root causes of gender violence and the value of peaceful conflict resolution techniques as alternatives to violence. Exploring systemic gender inequalities will contribute to the denormalization of gender-based violence.[47] These efforts should be implemented in primary and secondary schools, public and community-based campaigns, and at workplaces. All of these efforts will aid in the deconstruction of highly socialized gender identities while simultaneously championing a positive form of masculinity that avoids aggression or violence toward others.

---

[44] Oudraat et al., *Enhancing Security*, 18.
[45] Dotson and Frydman, *Neither Security nor Justice*, 15.
[46] Oudraat et al., *Enhancing Security*, 4, 18.
[47] Dotson and Frydman, *Neither Security nor Justice*, 14.

Conclusion

This chapter illustrates the endemic nature of gender-based violence. Gender-based violence sprouted from the same seeds that normalized violence, exacerbated structural inequalities, and increased impunity within the judicial system and the police force in postwar El Salvador. Postwar conditions also exacerbated the subjugation of women by undervaluing their rights, lives, and voices. While *machismo* informs and reproduces gender-based violence, the judiciary and police force legitimize, reinforce, and endorse it. In order to break the vicious cycle of subjugation that women endure, the judiciary and police force must be committed to increasing transparency, impartiality, and accountability. They must also be committed to incorporating gender perspectives into their strategic plans, policies, procedures, and operations. For these changes to be effective and long-lasting, civil society needs to systematically denormalize the use of violence as a form of conflict resolution in tandem with government efforts.

Examining gender-based violence through the lens of WPS illustrates that it is not simply a "women's issue." Tackling root causes of gender-based violence in El Salvador involves addressing a wide spectrum of broader issues that plague Salvadoran society: police and judiciary abuse of human and legal rights; judicial impunity; and how social norms can lead to discriminatory treatment of various groups in Salvadoran society. Ultimately, implementing the aforementioned policy recommendations will not only increase the protection and empowerment of half of El Salvador's population, but will also help Salvadoran society to purge itself of impunity and the normalization of violence.

PART 4

Professional Military Education

The Strategic Centrality of Women, Peace, and Security
A Call to Mainstream in Professional Military Education

by Lieutenant Colonel Casey M. Grider, USAF*

Codified in October 2000, UNSCR 1325, Women, Peace, and Security (WPS) has since become the banner beneath which marginalization of women and girls and gender-based harm, especially in international conflict, face indictment from a growing global community demanding equal participation, protection, and relief for this long-persecuted half of its population.[1] As we approach the 20-year anniversary of this historic declaration, the United States sits atop a mountain of policy, legislation, strategy, and plans. Like its namesakes, however, aspects of U.S. WPS activity have been marginalized and seldom see the same level of programmatic urgency as other, more traditional, central elements of national security and defense strategy.

Culturally institutionalized rights for women, meaningful participation in domestic and world affairs, and protection during and after armed conflict are foundational to achieving international peace and security. The United States must begin to demonstrably pursue its whole WPS strategy through deliberate implementation. The U.S. Air Force Air War College (AWC) is uniquely positioned to be a DOD pathfinder and must lead by migrating WPS philosophy from elective status into core AWC curricula on leadership, strategy, security, and warfighting. This chapter discusses the centrality of WPS to the successful pursuit of global security and surveys U.S. guidance and activity that over the past 20 years has, in many respects, broken barriers and blazed a trail for the world. It then turns a critical eye toward challenging WPS elements that have failed to gain traction and continue to await meaningful action. AWC is the best place to invigorate this critical institutional shift, and this chapter offers three practical examples of current courses where WPS can integrate as foundational strategic thought without disrupting or overburdening existing curricula. Finally, this chapter considers some common sources of hesitancy and even aversion to both WPS writ large and to mainstreaming implementation and offers counterpoints.

Building meaningful and sustainable international peace and security cannot succeed apart from considering the perspectives, needs, and welfare of entire populations, including their inherent gender, ethnic, social, and economic makeup. Extensive research highlights the especially critical importance of gender in this endeavor.

---

* The views expressed in this chapter are solely those of the author. They do not necessarily reflect the opinion of Marine Corps University, the U.S. Marine Corps, the U.S. Navy, the U.S. Army, U.S. Army War College, the U.S. Air Force, or the U.S. government.
[1] UN Security Council, Resolution 1325, Women, Peace and Security, S/RES/1325 (31 October 2000); and Australian Civil-Military Centre and UN Women, "Side by Side—Women, Peace and Security," YouTube, 18 June 2012, 0:38.

Considering the global economy and the impact of gender gaps, the McKinsey Global Institute mapped 15 gender equality indicators for 95 countries generating 97 percent of global GDP. They found that 40 of the 95 have extremely high or high levels of inequality in half or more of those indicators. If countries in 10 global regions could match the progress toward gender parity of their region's leader, global GDP would rise by as much as $12 trillion over 10 years.[2] Further, if the world were to reach the point where women contribute equally, its GDP would grow by as much as $28 trillion, equivalent to the 26 percent the United States and China generate together.[3] When the peace is shattered, however, these disparities move well beyond economics. Noting the disproportionate impact of violence during conflict, former U.S. Africa Command commander, retired Army general Carter Ham, noted, "In Africa's contemporary conflicts, more than 90 percent of all casualties are women and children, who also are more likely to be targets of sexual and gender-based violence."[4] Former U.S. Department of State assistant secretary of the Bureau of Conflict and Stabilization Operations, Rick Barton, said of conflict resolution, "Failing to include women in peace and security efforts results in a shaky, unstable, and partial peace that leaves in place a society's root causes of violence."[5] In his core mission to break cycles of violence and mitigate crisis, and he saw firsthand how stability is interwoven with the status of women and girls. Of 31 major peace processes between 1992 and 2011, 4 percent of signatories, 2.4 percent of chief mediators, 3.7 percent of witnesses, and 9 percent of negotiators were women.[6] During roughly the same period, however, where women were included at the table, peace agreements proved 20 percent more likely to last at least two years, and 35 percent more likely to last for 15 years.[7] Today's world is, therefore, forfeiting $28 trillion of its economic potential, losing significant portions of warzone female populations (and devastating future populations), and employing the perspective and skills of fewer than 10 percent of the demographic who have demonstrated the ability to influence a more lasting peace and security. Strategy simply cannot deliver desired end states when it fails to acknowledge or account for half the human terrain across the operational environment.

Ex nihilo nihil fit. Out of nothing, nothing comes. Women battled for empowerment long before UNSCR 1325. Just five years earlier, 17,000 participants from 189 coun-

---

[2] Jonathan Woetzel et al., *The Power of Parity: How Advancing Women's Equality Can Add $12 Trillion to Global Growth* (Shanghai: McKinsey Global Institute, 2015), 1.
[3] Woetzel et al., *The Power of Parity*, vii.
[4] "Women, War and Peace," UNIFEM.org, as cited in Carter Ham, "Working with African Nations to Support the Role of Women as Agents of Peace and Security," *Women on the Frontlines of Peace and Security* (Washington, DC: National Defense University Press, 2014), 114.
[5] Rick Barton and Cindy Y. Huang, "Creative Solutions for Crisis Response and Stabilization: The Power of a Gendered Approach," in *Women on the Frontlines of Peace and Security*, 25.
[6] Pablo Castillo Diaz et al., *Women's Participation in Peace Negotiations: Connections between Presence and Influence* (New York: UN Women, 2012), 2–3.
[7] Laurel Stone, "Annex II: Quantitative Analysis of Women's Participation in Peace Processes," in *Reimagining Peacemaking: Women's Roles in Peace Processes* (New York: International Peace Institute, 2015), 34.

tries and 30,000 activists produced the *Beijing Declaration and Platform for Action*, which formed a progressive and comprehensive blueprint "determined to advance the goals of equality, development, and peace for all women everywhere in the interest of all humanity."[8] In the years since, the world has seen significant gains in legal protection (274 legal and regulatory reforms in 131 countries), parity in education, and maternal mortality (down 38 percent), but far more work is yet to be done.[9] Novel to UNSCR 1325 is its focus on armed conflict. It specifically addresses military-related activities of peacekeeping; peace agreements; protection for women and girls from gender-based violence; gender-based war crimes and amnesty provisions; refugee camps and settlements; and disarmament, demobilization, and reintegration that considers female combatants.[10]

Resonating with this imperative, the United States has both exemplified and championed WPS. Two well-publicized examples involve combat. In 1991, the Senate voted to overturn a 43-year-old law barring women from flying warplanes in combat.[11] Soon after, then-lieutenant Jeannie Flynn (later, Leavitt) became the Air Force's first female fighter pilot, flying the McDonnell-Douglas F-15E Strike Eagle. Nineteen years later, she became the first woman to command an Air Force combat fighter wing.[12] In 2013, the DOD removed one of the last remaining barriers to women serving in the military by opening more than 14,000 positions previously restricted to men because they involved "direct combat on the ground."[13] The 17th chairman of the Joint Chiefs of Staff, Admiral Michael Mullin, captured the era's success perfectly: "Each time we open new doors in women's professional lives . . . we end up wondering why it took us so long."[14]

In addition to walking the walk and evolving to better promote women's meaningful participation, the United States has extensively codified its WPS commitment. Signed in 2011, President Barack Obama's Executive Order 13,595 directed the institution of both a NAP and Executive Agency Implementation Plans.[15] The *United States National Action Plan on Women, Peace, and Security* (WPS NAP) soon followed, laying out five interagency principles to guide WPS institutionalization, strengthen what existed, include women, and direct department and agency coordi-

---

[8] *Beijing Declaration and Platform of Action: Beijin +5 Political Declaration and Outcome* (New York: UN Women, 1995; repr., 2015), 2.
[9] *Gender Equality: Women's Rights in Review 25 Years after Beijing* (New York: UN Women, 2020), 4.
[10] UN Security Council, Resolution 1325, 2–3.
[11] Eric Schmitt, "Senate Votes to Remove Ban on Women as Combat Pilots," *New York Times*, 1 August 1991.
[12] Associated Press, "First Female Fighter Pilot Becomes First Female Wing Commander," 31 May 2012.
[13] David Vergun, "Secretary of Defense rescinds 'Direct Ground Combat Definition and Assignment Rule'," Army News Service, 24 January 2013; and Leon E. Panetta and Martin E. Dempsey, memo for secretaries of the military departments, acting under secretary of defense for personnel readiness, and chiefs of the military Services, "Elimination of the 1994 Direct Ground Combat Definition and Assignment Rule," 24 January 2013.
[14] Michael Mullen, "What Took Us So Long?: Expanding Opportunities for Women in the Military," in *Women on the Frontlines of Peace and Security*, 4.
[15] *Executive Order 13,595—Instituting a National Action Plan on Women, Peace, and Security* (Washington, DC: White House, 2011).

nation and accountability.[16] The then-president was calling upon his branch to adopt, collaborate, and follow through. In 2015, the *National Security Strategy* (NSS) deliberately stressed the need to protect women and girls in six passages, and called for increased women's participation across areas of conflict mediation, peace building, activism, economics, and politics.[17] Congress raised its voice in 2017, passing the Women, Peace, and Security Act, which codified in law U.S. support for WPS and directed a unified national strategy. Congress also went a step further, directing the DOD to train in areas of "conflict prevention, peace processes, mitigation, resolution, and security initiatives that *specifically address the importance of meaningful participation by women.*"[18] Today, the DOD boasts the fresh 2020 *Women, Peace, and Security Strategic Framework and Implementation Plan* (WPS SFIP), backed by a 2017 update to the NSS and a 2016 update to the WPS NAP that continue the drumbeat message that rights, meaningful participation, and protection are foundational to achieving international peace and security. To that end, the WPS SFIP establishes three objectives. First, the DOD must exemplify a diverse organization that allows for women's meaningful participation. Then, it must inspire partners to promote women's meaningful participation in defense. Finally, it must inspire partners to ensure women and girls are protected, especially during conflict and crisis.[19]

Broadly speaking, the United States' WPS guiding language tends to land in three categories: women's inclusion, women's protection, and wisdom. While the first two are self-evident, wisdom involves the difficult business of achieving gender awareness, integrating principles into existing frameworks, conducting informed strategic planning and execution, and possessing the cultural sensitivity to do it all with international partners. These are the challenging WPS elements that struggle to gain traction or see tangible results. Among the WPS SFIP's objectives, the latter two fit neatly into the inclusion and protection categories and face outward toward partner nations. Remaining on the table is a call to wisdom.

It takes wisdom to learn from missteps, faithfully uncover root causes and contributing factors, and internalize lessons as an organization. On 23 March 2003, Iraqi forces attacked a convoy and took prisoners of war, including Private Jessica Lynch of the U.S. Army's 507th Maintenance Company. She was held for a week and suffered sexual assault.[20] Inaccurate public portrayal and media attention quickly sent the narrative spiraling and harmed DOD's image. A year later, on 28 April 2004, photographs of female U.S. soldiers humiliating Abu Ghraib prison detainees sur-

---

[16] *United States National Action Plan on Women, Peace, and Security* (Washington, DC: Office of the President, White House, 2011), 1.
[17] National Security Strategy (Washington, DC: Office of the President, White House, 2015), 10–11.
[18] Women, Peace, and Security Act of 2017, Pub. L. No. 115-68 (2017), emphasis added.
[19] *Women, Peace, and Security Strategic Framework and Implementation Plan* (Washington, DC: U.S. Department of Defense, 2020), 7.
[20] Joana Cook, *A Woman's Place: U.S. Counterterrorism Since 9/11* (London: Hurst Publishers, 2018), 158–59.

faced, igniting a call to arms for recruiting insurgents to fight American forces in Iraq. Former president George W. Bush would later call it "the biggest mistake" made by the United States in Iraq. To make matters worse internally, Army Reserve brigadier general Janis Karpinski would also emerge as an example of poor leadership and refusal to establish and enforce standards.[21] It also takes wisdom to correctly identify the main priorities. Retired colonel Sheila Scanlon, Afghanistan's first appointed gender advisor, saw firsthand the difficulty of applying the NAP in the field. She operated without guidance and noted, "Between 2012–14, I never knew there was a NAP and that it had been signed in 2013. Nobody ever talked about it or mentioned it."[22] In an ad hoc position with no directive, action plan, policy, or strategic guidance, she was severely limited in making WPS relevant. General David H. Petraeus would note that policy and directives focused on enabling Afghan forces to secure their country, and he had limited resources to do so. Though the notion of women and their status was clearly important, it was not the core objective.[23] One can argue that any objective of any substance will inevitably fall prey to conflicting priorities unless it is considered indispensable to the core mission.

The DOD, therefore, requires wisdom to discern where WPS impacts strategy. The DOD's WPS SFIP calls for the development of doctrine, training, and education that reflects WPS principles. Specifically, it seeks increased awareness and an ability to integrate into missions.[24] U.S. Air Force AWC sequesters DOD's rising strategic leaders for one year to reflect on the past, analyze themselves, and form ideas about how they, as future strategy makers and strategic leaders, will steward the department through the next decade. As they navigate AWC's comprehensive curricula, WPS thought applies throughout. AWC students also learn side-by-side with international partner students, offering a prime opportunity to prioritize the NAP and WPS SFIP's objectives of being the exemplar, inspiring promotion of women's meaningful participation, and inspiring the protection of women and girls.

Recommendations and Conclusions

To begin blazing a deliberate academic trail, core AWC courses are already primed to incorporate WPS concepts. In the Foundations of Strategy course, Instructional Period 6427 focuses on the Iraq War. A third desired learning outcome (DLO) might be: "Analyze the impact of women's contributions in Iraq from 2003–06." One added question for study and discussion might be: "In what ways were Servicewomen's involvement in Iraq beneficial, controversial, or damaging? Consider the strategic impact of Team Lioness, Private Jessica Lynch, and the events of Abu Ghraib." These

---

[21] Cook, *A Woman's Place*, 160–61.
[22] Cook, *A Woman's Place*, 197.
[23] Cook, *A Woman's Place*, 181–82.
[24] *Women, Peace, and Security Strategic Framework and Implementation Plan*, 16.

could be supported by one 15-page reading from Dr. Joana Cook's book *A Woman's Place: U.S. Counterterrorism Since 9/11*.[25]

Next, in the Strategic Leadership and the Profession of Arms course, Instructional Period 6204 addresses the concepts of vision and leading change. A third DLO might be: "Analyze the impact of WPS on the DOD's vision and how progress for women in the United States ranks and influences relationships with international partners." One added question for study and discussion might be: "How might a senior leader approach gender concerns in an organization and cast a vision that maximizes the potential mission impact of all unit members?" These could be supported by a six-page reading from retired admiral William McRaven in *Women on the Frontlines of Peace and Security*, along with the DOD WPS SFIP for policy context.[26] Finally, the AU Commandant's Lecture Series could deliberately feature a woman who has shown successful strategic military leadership, showcasing some of her thoughts in the WPS arena. Some potential examples might include retired Air Force general Lori Robinson, a former U.S. Northern Command commander; Air Force major general Jeannie Leavitt, Headquarters Air Education and Training Command's current director of operations and communications; or retired Army general Ann Dunwoody, the first female four-star general.

Leading change is difficult, but this is not headline news at AWC. In a fine-tuned machine designed to optimally balance all that can be squeezed into one academic year, capacity to incorporate another subject claiming to be foundational is limited. While this presents a real curriculum development challenge, the question should not be whether to shoehorn in additional material. If the faculty is open to comparing WPS to other core concepts, its relevance to half of the world's population and fundamental influence on strategic outcomes will make it a competitive candidate. This area of study is also enormous in its own right. It would be foolhardy to believe all the material could simply migrate to the middle. That, however, should not be the goal. A simple focus of raising awareness and applying thought where it is strategically relevant would suffice. The WPS elective course for deeper study should not cease to exist but rather grow in enrollment.

Twenty years after UNSCR 1325 breathed life into WPS principles, the United States has accomplished much, but has much still to accomplish. It will always be of central importance to consider equal rights, meaningful participation, and equal protection in all aspects of both domestic and international affairs. The United States' WPS activity has not yet risen to its rightful place in our collective consciousness, but the opportunities to push that direction are clear and practical. AWC has a unique

---

[25] Cook, *A Woman's Place*, 149–64.
[26] William H. McRaven, "Women in Special Operations Forces: Advancing Peace and Security through Broader Cultural Knowledge," in *Women on the Frontlines of Peace and Security* (Washington, DC: National Defense University Press, 2014), 127–34; and *Women, Peace, and Security Strategic Framework and Implementation Plan*.

opportunity to push WPS philosophy into core strategic thought and lead the United States toward full actualization of its own espoused national identity and strategy.

This chapter has argued for the centrality of these concepts. A survey of developing thought and guidance showed where we have been and where we should be headed. It focused on the particular difficulty of developing WPS-informed wisdom and how that might take root in Air University. Finally, it attempted to scratch the surface of resistance inherent to any endeavor to push organizational change, and it offered counterpoints to reinforce the truth that implementation is both practical and essential.

PART 5

Vietnam

The Nexus of Climate Change, Migration, and Human Trafficking

by Ms. Amy Patel*

When human trafficking researchers interviewed Vietnamese migrants at a makeshift camp in Europe in 2017 to find out why the number of migrants and trafficking victims had swelled that year, they were surprised to hear environmental changes were a top reason.[1] A toxic waste spill along Vietnam's coast had wiped out the local fishing industry and motivated thousands to search for work overseas. Many became victims of human smugglers and traffickers. Similar environmental degradation from climate change is likely to accelerate migration, human smuggling, and human trafficking of Vietnamese in the decades ahead. Research shows that women make up the majority of migrants in Vietnam and are at particular risk of gender-based violence and human trafficking.[2]

This paper uses Vietnam as a case study to examine the effects of climate change; the relationship between climate change, migration, human smuggling, and human trafficking; and the specific risks for women. It concludes with a discussion of actions that Vietnam and the international community can take to manage the impact of climate change, particularly on women. If Vietnam's government is willing to prioritize sustainable development and gender-inclusive decisionmaking, there are steps it can take to mitigate the effects of climate change, support migrants, protect trafficking victims, and reduce human trafficking.

The Effects of Climate Change in Vietnam

The World Bank ranks Vietnam among the top five countries most affected by climate change.[3] The country's long, heavily populated coastlines and river deltas are vulnerable to sea rise and increasingly frequent extreme weather events.[4] The latest projections indicate that by 2050, sea levels could rise high enough to flood the land where one-quarter to one-third of the population lives today.[5] Man-made activities such as

---

* The views expressed in this chapter are solely those of the author. They do not necessarily reflect the opinion of Marine Corps University, the U.S. Marine Corps, the U.S. Navy, the U.S. Army, U.S. Army War College, the U.S. Air Force, or the U.S. government.

[1] *Precarious Journeys: Mapping Vulnerabilities of Victims of Trafficking from Vietnam to Europe* (London: ECPAT UK, 2019), 36–37, 45; and James Pearson, "Postcards from a Poisoned Coast: Vietnam's People-Smuggling Heartland," Reuters, 28 October 2019.

[2] *Migration, Resettlement and Climate Change in Viet Nam* (Hanoi: United Nations in Vietnam, 2014), 2.

[3] Jun Rentschler et al., *Resilient Shores: Vietnam's Coastal Development Between Opportunity and Disaster Risk* (Washington, DC: World Bank, 2020), 10.

[4] Fiona Miller and Olivia Dun, "Resettlement and the Environment in Vietnam: Implications for Climate Change Adaptation Planning," *Asia Pacific Viewpoint* 60, no. 2 (August 2019): 133, https://doi.org/10.1111/apv.12228; and Amit Prakash, "Boiling Point," *Finance & Development Magazine* 55, no. 3 (September 2018): 22.

[5] Rentschler et al., *Resilient Shores*, 10; "Flooded Future: Global Vulnerability to Sea Level Rise Worse than Previously Understood," Climate Central, 29 October 2019; and Denise Lu and Christopher Flavelle, "Rising Seas Will Erase More Cities by 2050, New Research Shows," *New York Times*, 29 October 2019.

sand mining and upstream hydropower dams "pose greater threats in the short term while exacerbating medium- and long-term climate change impacts."[6]

The regions of Vietnam most likely to suffer from climate change are the same areas that produce most of the country's food. One study predicts that sea level rise will affect seven percent of Vietnam's agricultural land within the next 30 years.[7] Sea level rise, more frequent drought, saltwater intrusion, and other effects of climate change will reduce the agricultural productivity of the Mekong delta, which produces more than half of Vietnam's rice and 60 percent of its shrimp.[8] Research shows that "women play a central role in agriculture in Vietnam, contributing more hours of labor than men in cultivation, livestock breeding, agricultural processing and agriculture produce marketing."[9] This makes women more vulnerable to the effects of extreme weather changes and natural disasters.

Because many of the safest areas in Vietnam are already heavily developed, a disproportionate share of new development is happening along the coast and in low-lying areas at highest risk of climate change. The regions fueling Vietnam's economic growth are the most vulnerable to disasters—a recipe for humanitarian catastrophes and economic losses in the future.[10]

Destroyed ports, aquaculture farms, and tourism infrastructure would exacerbate the economic damage and "shrink local tax bases, straining municipalities' abilities to pay for public goods such as education."[11] Public infrastructure is also at risk; the World Bank estimates that 22 percent of schools and 26 percent of health care facilities in coastal provinces are exposed to intense floods.[12] Climate change could drive hundreds of thousands into poverty and reverse years of economic gains and poverty reduction.

The Link Between Climate Change and Migration

Sea rise, increasingly frequent natural disasters, and other effects of climate change are likely to increase migration. "Climate change and environmental disasters that destroy livelihoods are push factors" for people to migrate internally, which is often accompanied by increased migration abroad.[13] Studies indicate that climate change is already starting to drive migration from the Mekong delta provinces. Migration from areas most vulnerable to climate change is more than double the national aver-

---

[6] Sen Nguyen, "Chinese Dams, Pollution Send Vietnamese in Mekong Delta in Search of Greener Pastures," *South China Morning Post*, 28 February 2021.
[7] Rentschler et al., *Resilient Shores*, 10.
[8] Prakash, "Boiling Point," 25.
[9] Rentschler et al., *Resilient Shores*, 41.
[10] Rentschler et al., *Resilient Shores*, 21.
[11] "Flooded Future"; and Prakash, "Boiling Point," 22.
[12] Rentschler et al., *Resilient Shores*, 18.
[13] *Precarious Journeys*, 45.

age.[14] One study found that "climate change is the dominant factor in the decisions of 14.5 percent of migrants leaving the Mekong delta."[15]

Some well-intentioned climate change mitigation measures have inadvertently increased migration. For example, thousands of miles of new dykes in the Mekong delta have prevented flooding of homes, but they also altered the ecosystem, killed fish, and sapped the land of nutrients. As a result, farmers and fishers are no longer able to support their families, and they are moving out of the delta.[16]

In other cases, migration is the result of planned relocations sponsored by the government to protect communities from climate change. "The Vietnamese government considers resettlement to be amongst its key strategies for climate change adaptation."[17] Vietnam has improved land rights, compensation, and legal protections for relocated households, but it still lacks the capacity at the local level to ensure long-lasting, positive outcomes for resettled families.[18] If migrants cannot quickly recreate social networks and income streams, many end up worse off after relocation and slip further into poverty.

Women in Vietnam are disproportionately affected by migration and its consequences. Research indicates that "women make up the majority of migrants" in Vietnam due to the high demand for domestic servants and female workers in industrial zones.[19] Moreover, "female migrant factory workers may be subject to gender-based violence, from their partners as well as from individuals in the community as they are living away from the protection of their families and lack social networks in the migrant housing areas."[20]

More Migration Means More Smuggling and Trafficking Victims

Migrants displaced by the effects of climate change are vulnerable to exploitation by human smugglers and human traffickers. When internal migration within Vietnam increases, it is often accompanied by a corresponding rise in migration abroad in

---

[14] Alex Chapman and Van Pham Dang Tri, "Climate Change Is Triggering a Migrant Crisis in Vietnam," *The Conversation*, 9 January 2018.
[15] Chapman and Tri, "Climate Change Is Triggering a Migrant Crisis in Vietnam"; Nguyen, "Chinese Dams, Pollution Send Vietnamese in Mekong Delta in Search of Greener Pastures"; and Oanh Le Thi Kim and Truong Le Minh, "Correlation between Climate Change Impacts and Migration Decisions in Vietnamese Mekong Delta," *International Journal of Innovative Science, Engineering & Technology* 4, no. 8 (August 2017): 111–18.
[16] Chapman and Tri, "Climate Change Is Triggering a Migrant Crisis in Vietnam"; Joshua Kurlantzick, "In the Face of Catastrophic Sea Level Rise, Countries in Southeast Asia Dither," *World Politics Review*, 11 November 2019; and Kim and Minh, "Correlation between Climate Change and Migration Decisions in Vietnamese Mekong Delta."
[17] Miller and Dun, "Resettlement and the Environment in Vietnam," 133.
[18] Miller and Dun, "Resettlement and the Environment in Vietnam," 138.
[19] *Migration, Resettlement and Climate Change in Viet Nam*, 2.
[20] *Migration, Resettlement and Climate Change in Viet Nam*, 2.

search of better job opportunities and living conditions.[21] If legal pathways are closed off or difficult to navigate, migrants turn to human smugglers who often exploit them on the journey abroad.

Although migration, human smuggling, and human trafficking are distinct concepts, they are often conflated. "Migrants tend to take illicit and dangerous routes, making them easy prey for criminal networks."[22] Researchers interviewing Vietnamese trafficking victims in Europe found that many began their trips abroad as an attempt to seek legitimate employment, but they later discovered they were part of a human smuggling ring. Finding themselves in a vulnerable position, they were at risk of exploitation and became trafficking victims along the journey through Europe.[23] "Women and girls are particularly susceptible to sexual exploitation at the hands of traffickers and even the peers they are travelling with."[24]

The U.S. Department of State's "2020 Trafficking in Persons Report: Vietnam," placed Vietnam on its Tier 2 Watch List because the country "does not fully meet the minimum standards for the elimination of trafficking."[25] This designation puts Vietnam at risk of losing U.S. assistance if it does not improve its record, according to the policies set forth in the Trafficking Victims Protection Act of 2000.[26] The report notes that trafficking victims include women and children who travel abroad for internationally brokered marriage or for jobs where they are subjected to forced labor in domestic service or sex trafficking. The report concludes that Vietnam is "making significant efforts" to improve its response to human trafficking; however, coordination between national and provincial level officials is poor.[27] The UK government notes that Vietnam's national action plan on antitrafficking does not address the link between labor migration and trafficking.[28]

Experts have identified environmental factors as one of the root causes of human trafficking. As climate change increases the frequency and intensity of natural disasters, the flow of potential trafficking victims will grow.[29] According to the UN Environ-

---

[21] Precarious Journeys, 45.
[22] Mely Caballero-Anthony, "A Hidden Scourge," Finance & Development Magazine 55, no. 3 (September 2018): 20.
[23] Precarious Journeys, 11.
[24] Precarious Journeys, 65.
[25] "2020 Trafficking in Persons Report: Vietnam," U.S. Department of State, June 2020. The U.S. Department of State ranks countries based on a four-tier system, as required by the Trafficking Victims Protection Act (TVPA) of 2000. The tiers rank the extent of a government's efforts to meet the TVPA's minimum standards for eliminating human trafficking. Tier 1 denotes governments that fully meet the minimum standards; Tier 2 governments do not fully meet the standard but are making significant efforts; Tier 2 Watch List governments do not fully meet the standard and their countries' rate of trafficking is very significant or increasing; and Tier 3 governments are making no efforts to meet the minimum standard. Trafficking in Persons Report, 20th ed. (Washington, DC: U.S. Department of State, 2020), 39–41.
[26] Trafficking in Persons Report, 42.
[27] "2020 Trafficking in Persons Report: Vietnam."
[28] Precarious Journeys, 33.
[29] Precarious Journeys, 38, 45.

ment Programme, "trafficking may increase 20–30 percent during disasters."[30] For example, after Typhoon Haiyan in the Philippines in 2013, researchers documented an increase in trafficking victims forced to work as domestic servants, beggars, prostitutes, and laborers.[31]

The 2016 toxic waste spill at a steel factory polluted more than 100 miles of Vietnam's coastline. The environmental disaster ruined the local fishing and tourism industries, resulting in "a dramatic increase in the number of regular and irregular Vietnamese migrants and trafficking victims overseas."[32] Vietnamese at a migrant camp in France cited the job losses after the toxic waste spill as the reason they had left.[33] Although this was a man-made disaster, climate change will produce similar environmental degradation and economic losses that trigger migration. Whether migrating internationally or internally within Vietnam, migrants are vulnerable to traffickers who promise job opportunities.

The relationship between human trafficking and environmental degradation also works in reverse. Trafficking victims provide extremely cheap labor "that has been shown to contribute to deforestation and to highly polluting methods of shrimp farming."[34] Carbon emissions from deforestation are a leading source of greenhouse gases.[35] This cycle is reinforcing: the effects of climate change increase the number of trafficking victims, and the growing supply of cheap trafficked labor increases the profitability of industries that exacerbate climate change.

What to Do?

Vietnam can take many steps to prevent human trafficking, protect trafficking victims, support migrants, and reduce migration from climate change. The United States and other international partners are well-placed to support these efforts, and Vietnam has welcomed many cooperative projects already to address climate change, migration, and human trafficking.

Address Human Trafficking

First, border guards and maritime authorities need training and financial incentives to detect trafficking, assist victims, and arrest traffickers. Stronger enforcement of trafficking laws and stiffer penalties would increase the traffickers' cost of doing business.[36] Efforts to combat associated crimes—such as smuggling, prostitution, organ

---

[30] Michael B. Gerrard, "Climate Change and Human Trafficking after the Paris Agreement," *University of Miami Law Review* 72, no. 2 (Winter 2018): 358.
[31] Caballero-Anthony, "A Hidden Scourge," 20.
[32] *Precarious Journeys*, 44.
[33] *Precarious Journeys*, 45.
[34] Gerrard, "Climate Change and Human Trafficking after the Paris Agreement," 359.
[35] Prakash, "Boiling Point," 25.
[36] Gerrard, "Climate Change and Human Trafficking after the Paris Agreement," 368; and *Transnational Organized Crime in Southeast Asia: Evolution, Growth and Impact* (Bangkok, Thailand: United Nations Office on Drugs and Crime, 2019), 83.

trafficking, and money laundering—would reduce the demand for trafficked labor and make migration safer.[37]

Vietnam should also improve its identification and protection of trafficking victims. The Department of State's "2020 Trafficking in Persons Report: Vietnam" notes that "the government had common victim identification criteria as part of the Coordinated Mekong Ministerial Initiative against Human Trafficking and its own 2014 procedure for victim identification; however, neither the criteria nor the procedures were reported to be proactively or widely employed, including among women arrested for commercial sex acts."[38] In addition, Vietnam's government and nongovernmental organizations should raise awareness about trafficking within communities at higher risk of experiencing climate change. Greater awareness would help ensure migrants do not fall victim to traffickers' false promises.[39]

Reduce Climate-Induced Migration

Actions to prevent migration would reduce the number of people vulnerable to trafficking. Vietnam should help communities adapt to climate change and prepare for environmental disasters so they are more resilient and will not have to migrate.[40] In addition, new development projects must consider the likely effects of climate change. This requires better information about the impact of climate change and risk-informed planning for new infrastructure projects. "If Vietnam does not consider the impacts of natural hazards and climate change in its investment and planning decisions, the number of people falling into poverty will pose a serious threat to its ambitious goals for poverty reduction."[41]

Smarter Resettlement

When resettlement is required, there are several steps Vietnam can take to plan for more orderly relocations, to support migrants, and to learn from its previous mistakes with resettlement.[42] Experts recommend involving migrants in decisionmaking about their resettlement, maintaining social networks in the new location, and connecting migrants with new jobs. The inclusion of women in decisionmaking about resettlement programs is particularly important to ensure these efforts are gender-responsive and address the factors that make women vulnerable to gender-based violence and human trafficking.

---

[37] Caballero-Anthony, "A Hidden Scourge," 20.
[38] "2020 Trafficking in Persons Report: Vietnam."
[39] *Transnational Organized Crime in Southeast Asia*, 83.
[40] Gerrard, "Climate Change and Human Trafficking after the Paris Agreement," 367.
[41] Rentschler et al., *Resilient Shores*, 36.
[42] Lu and Flavelle, "Rising Seas Will Erase More Cities by 2050, New Research Shows"; Miller and Dun, "Resettlement and the Environment in Vietnam," 143; and Gerrard, "Climate Change and Human Trafficking after the Paris Agreement," 368.

Research indicates Vietnamese resettlement efforts "often lack in-depth gender analysis, which is important in climate change adaptation, disaster risk reduction, and resettlement." Experts recommend that "women's empowerment and participation in decision-making processes, including those related to resettlement, should be promoted as an integral part of resettlement processes."[43] These steps considerably increase the prospects for successful outcomes after climate change-induced relocations.[44]

Conclusion

Vietnam's government recognizes the threat of climate change and has been working with international partners, including the United States. These efforts are mutually beneficial and should continue. Nevertheless, Vietnam's sustainable development goals often conflict with policies to support economic development, such as electricity subsidies and construction of coal-fired power plants.[45] As one expert on Southeast Asia has noted, "countries like Vietnam, where governments are autocratic and rely on economic development to maintain their legitimacy, may be especially loath to undermine economic growth."[46] Ultimately, the Vietnam government's ability to navigate this tension will determine whether it succeeds in mitigating the effects of climate change, including migration, human smuggling, and human trafficking. Such efforts will have a greater possibility of success if they incorporate an analysis of gender dynamics and include women in the decision-making process.

---

[43] *Migration, Resettlement and Climate Change in Viet Nam*, 25.
[44] Miller and Dun, "Resettlement and the Environment in Vietnam," 138, 143.
[45] Prakash, "Boiling Point," 26.
[46] Kurlantzick, "In the Face of Catastrophic Sea Level Rise, Countries in Southeast Asia Dither."

PART 6

Hegemonic Masculinity

The Effect of Hegemonic Masculinities on the Endemic of Sexual Misconduct in the U.S. Army

by Major Sarah E. Salvo, U.S. Army[*]

Chapter 1
Introduction

Sexual assault and harassment have been a cancer within the U.S. military for decades. DOD and Congress have struggled to adjudicate the problem for years with very little success. Congressional officials have charged the U.S. armed forces of perpetuating a "rape culture," yet very little literature exists that actually examines what rape culture looks like within the U.S. military. Further, the preponderance of offenders of sexual harassment and assault within the military are men. However, American society and military and government leaders continue to label sexual harassment and assault predominately a women's issue rather than seeking to understand the role of men and masculinity in perpetuating rape culture.[1]

This study seeks to understand the relationship between hegemonic masculinity and sexual harassment and assault within the U.S. Army's organizational culture. Additionally, this thesis examines formal and informal organizational culture to understand what aspects of Army culture may be preventing leaders from seeing the signs and symptoms of sexual harassment and assault. With a thorough understanding of underlying aspects of Army culture that create opportunities for sexual harassment and assault to occur, Army leaders can be armed with the knowledge to effect positive cultural change that is long overdue.

For the last 30 years, sexual harassment and assault scandals within the U.S. Army have persisted despite the establishment of formalized programs to prevent its occurrence. In 1988, the DOD conducted the first *Survey on Sex Roles in the Active-Duty Military* prompted by the U.S. Merit System Protections Board identification of large-scale sexual harassment occurring within the public sector and government.[2] The survey estimated that upwards of 22 percent of active-duty military personnel (64 percent of women and 17 percent of men) reported one or more incidents of unwanted, uninvited sexual attention in the workplace.[3] These appalling figures drew outrage among the

---

[*] The views expressed in this chapter are solely those of the author. They do not necessarily reflect the opinion of Marine Corps University, the U.S. Marine Corps, the U.S. Navy, the U.S. Army, U.S. Army War College, the U.S. Air Force, or the U.S. government.
[1] Guy Raz, interview with Jackson Katz, "Why We Can No Longer See Sexual Violence as a Women's Issue," transcript, TED Radio Hour, National Public Radio, 1 February 2019.
[2] John B. Pryor, *Sexual Harassment in the United States Military: The Development of the DOD Survey* (Normal, IL: Defense Equal Opportunity Management Institute, 1988), 1.
[3] Lisa D. Bastian, Anita R. Lancaster, and Heidi E. Reyst, *Department of Defense 1995 Sexual Harassment Survey* (Arlington, VA: Defense Manpower Data Center, 1996).

public and lawmakers, prompting DOD to create the Equal Employment Opportunity (EEO) Program with the charge of prevention of workplace sexual harassment.

In 1996, two major sexual harassment and assault scandals within the U.S. Army unraveled, surrounding Sergeant Major of the Army Gene C. McKinney, and advanced individual training (AIT) instructors at Aberdeen Proving Grounds (APG).[4] The scandals followed formal sexual harassment complaints filed by a female trainee against an APG instructor. Following the initial report, approximately 34 women came forward to file sexual harassment and assault reports against APG instructors. The incidents prompted the U.S. Army to set up a sexual assault hotline, which would soon receive more than 6,000 calls alleging widespread abuse across the Army.[5]

The Army immediately launched internal investigations into these incidents, focusing on the events occurring at APG. During the investigation, what became known as the "The Game" scandal was uncovered. "The Game" was a competition created by AIT leaders to see who among the leadership could have sex with the most trainees.[6] Amid the scandal, the newly appointed Sergeant Major of the Army (SMA) McKinney visited installations around the Army urging soldiers that the equal opportunity system would work but that in order for it to work, soldiers should come forward and report claims of abuse.[7] McKinney soon became exposed as a perpetrator of sexual harassment and assault as six women filed reports claiming he made unwanted sexual advances toward them on multiple occasions.[8]

The Army ultimately punished AIT leaders at APG and McKinney, however in comparison to the gravity of offenses committed by these perpetrators, the consequences were mild. Of the three individuals involved in "The Game," one received a 25-year military prison sentence and the others received six to four months in military prison. McKinney, the most senior non-commissioned officer in the U.S. Army, faced no criminal charges for his misconduct.[9] The court system demoted Sergeant Major of the Army McKinney to the rank of master sergeant (E8), issued him a letter of reprimand for obstructing justice, and permitted him to retire.[10] The reduction did not stand, as it violated U.S. Code, so McKinney retired as a sergeant major (E9), allowing the collection of a pension at his current rank.[11]

These incidents were clear indicators that the Army had a severe and widespread problem with sexual harassment and assault, especially considering that many of the perpetrators were male leaders with positions of trust and confidence. These incidents occurred almost eight years after establishing the EEO Program, the

---

[4] Joseph E. Webster, "Resisting Change: Toxic Masculinity in the Post Modern United States Armed Forces, (1980s–present)" (PhD diss., University of Central Oklahoma, 2019).
[5] Webster, "Resisting Change," 51.
[6] Webster, "Resisting Change," 52.
[7] Webster, "Resisting Change," 54.
[8] Webster, "Resisting Change," 54.
[9] Webster, "Resisting Change," 53.
[10] Webster, "Resisting Change," 59.
[11] Webster, "Resisting Change," 57.

sole focus of which was conducting organizational training to prevent sexual harassment and assault in the workplace. Lastly, the judicial system undermined the likelihood of victim reporting, as offenders of sexual harassment and assault continued to receive mild to no punishments for grave abuses of power and ranks that violate all Army values.

Instead of focusing on the cultural issues leading to the abhorrent behavior occurring within the Army, major conflicts captured the attention of Army and congressional leadership from the late 1990s to present date. U.S. involvement with conflicts in Bosnia and Herzegovina, the attacks of 11 September 2001, and the beginning of the Iraq and Afghanistan conflicts dominated the narrative and the Army's focus. However, like any cancerous behavior within an organization, it did not take very long for the systemic occurrence of sexual harassment and assault to resurface.

In 2004, several years after the start of conflicts in both Iraq and Afghanistan, servicemembers began reporting sexual abuses occurring in combat. The reporting of sexual harassment and assault significantly increased, prompting media reports and deep criticisms that military leadership was not taking the misconduct seriously.[12] In response, Secretary of Defense Donald Rumsfeld formally established the Sexual Assault Prevention and Response (SAPR) Program with the primary mission of tracking reports of sexual assaults, supporting victims with medical attention, counseling, and reporting options, and conducting sexual assault prevention training.[13] Before the SAPR Program establishment, victims of sexual assault had no medical care resources or mental health support. Subsequently, congressional mandates to report sexual assault began in calendar year 2004 and became an annual requirement in the 2005 National Defense Authorization Act (NDAA).[14] It appeared the DOD was finally trending in the right direction to address sexual harassment and assault by all accounts.

Shortly after the U.S. Army established the SAPR Program, Chief of Staff of the Army General Raymond Odierno directed the SAPR Program to reorganize, absorbing the Military Prevention of Sexual Harassment training responsibility formally owned by the EEO Program. General Odierno recognized that sexual harassment and sexual assaults were not happening independently of one another.[15] The reorganization resulted in the creation of the Sexual Harassment and Assault Response Prevention (SHARP) Program, known today as the proponent of sexual harassment and assault prevention and response in the U.S. Army. The program's goal remained the same but simply added the responsibility of providing victims of sexual harassment the same support and reporting options available to victims of assault.[16]

---

[12] Angela Andrew, *Leading Change: Sexual Harassment/Assault Response and Prevention (SHARP)* (Carlisle, PA: U.S. Army War College, 2013), 2.
[13] Andrew, *Leading Change*, 4.
[14] Andrew, *Leading Change*, 3.
[15] Andrew, *Leading Change*, 5.
[16] Andrew, *Leading Change*, 6.

As conflict raged on in the Middle East, reports of sexual harassment and assault continued to rise as the Army integrated the SHARP Program. Reports from 2004 through 2011 more than doubled, rising from 725 to 1,695.[17] Army leaders quickly justified the reporting increase as the program working and more and more victims being confident in reporting sexual harassment and assault. This rationalization soon was proved inaccurate as major flaws within the SHARP Program became magnified.

In 2014, Sergeant First Class Gregory McQueen, a victim advocate for the SHARP Program in Fort Hood, Texas, was exposed for organizing a prostitution ring consisting of female junior enlisted soldiers under his command. McQueen essentially groomed the subordinate soldiers to participate in the ring by promising them they could make serious money at the parties he was organizing. McQueen organized the parties for senior officers to have sex with the women. McQueen rented out hotel rooms where the prostituted soldiers would meet higher-ranking officials for paid sex. Additionally, McQueen hosted parties where he put the soldiers on display and pimped them out to attendees, who were senior officers. McQueen was court-martialed on multiple charges, which amounted to 40 years in prison. Instead, McQueen pleaded guilty, resulting in his actual punishment being much less severe. He was reduced to private (E1), sentenced to two years in prison, and given a dishonorable discharge from the Army.[18] Additionally, few details on the consequences of the senior officers known to have frequented these parties are publicly available.

The typical public, DOD, and congressional reactions followed McQueen's court-martial. The event triggered major changes to the screening of victim advocates within the SHARP Program, requiring more stringent training standards, rank requirements, and background checks for program appointees. The 2014 Fort Hood Prostitution Ring Scandal drew further scrutiny to the SHARP Program's effectiveness and the U.S. Army's ability to effectively deliver justice to victims. Following this incident and rising statistics across DOD, Senator Kirsten Gillibrand, a member of the Senate Armed Forces Committee, began introducing legislation to remove the prosecution of sexual harassment and assault from military commanders' discretion.[19]

Following the 2014 Fort Hood Prostitution Ring Scandal, congressional inquiry into sexual assault within the U.S. Army became more frequent. From 2008 through 2019, the U.S. Army reports of sexual assault saw an upward trend with slight variances between years (table 6.1).

---

[17] Andrew, *Leading Change*, 3.
[18] M. L. Nestel, "Inside Fort Hood's Prostitution Ring," *Daily Beast*, 14 April 2017.
[19] "Military Justice Improvement Act: Supportive Editorials," Kirsten Gillibrand, United States Senator for New York, accessed 13 June 2019.

Table 6.1. Army Sexual Assault Reports by Year

Reports of sexual assaults (rate/1,000)	FY 2008	2009	2010	2011
Unrestricted reports[1]	1,476	1,658	1,482	1,520
Restricted reports	256	283	299	301
Total reports[1]	1,732	1,941	1,781	1,821
Total servicemember victims[2]	1,337	1,397	1,316	1,378
Servicemember report rate/1000[3]	2.5	2.6	2.4	2.5

Reports of sexual assaults (rate/1,000)	FY 2012	2013	2014	2015
Unrestricted reports[1]	1,398	2,017	2,199	2,046
Restricted reports	174	318	407	470
Total reports[1]	1,572	2,335	2,606	2,516
Total servicemember victims[2]	1,248	1,766	2,072	1,922
Servicemember report rate/1000[3]	2.3	3.5	4.2	4.2

Reports of sexual assaults (rate/1,000)	FY 2016	2017	2018	2019
Unrestricted reports[1]	1,996	2,178	2,576	2,551
Restricted reports	501	528	579	668
Total reports[1]	2,497	2,706	3,155	3,219
Total servicemember victims[2]	1,962	2,123	2,501	2,536
Servicemember report rate/1000[3]	4.4	4.7	5.5	5.5

1: As of FY14, one victim equals one report, per DOD guidance. (FY08–FY13 adjusted to one victim per report).
2: Includes only servicemember victims in restricted and unrestricted reports for incidents occuring while in the military.
3: Includes servicemembers reporting incidents occurring prior to military service.

Source: *Department of Defense Annual Report on Sexual Assault in the Military, Fiscal Year 2019* (Washington, DC: U.S. Department of Defense, 2020), Enclosure 1: Department of the Army, 28.

Regarding sexual harassment, the DOD has collected top-line estimates of the incidence of sexual harassment in the workplace since 1988 and quantifies the number of formal reports received annually across the services. Formal reports of sexual harassment collected by the Army and DOD gender relations survey data on sexual harassment have significant disparities. The 2018 and 2019 *DOD Gender and Workplace Relations Survey for Active Duty Forces* estimated sexual harassment rates of 6.3 percent for men and 24.2 percent for women, while only 1,021 formal sexual harassment complaints were filed across the entire DOD. Further, DOD estimates that only 1 in 3 servicemembers report sexual harassment and assault to a DOD authority.[20] This disparity is cause for speculation that the occurrence of sexual harassment and assault is far greater than DOD estimates and perhaps may not have changed much from the first DOD survey in 1988, which estimated that upwards of 60 percent of women and 20 percent of men experienced some form of sexual harassment in the workplace.[21]

The events of 2020 have been a clear example that the Army's current approach to addressing sexual harassment and assault is not working and that its occurrence is indeed much more widespread than survey data estimates. In April 2020, Specialist Vanessa Guillén disappeared from Fort Hood without a trace. Guillén was a hardworking, dedicated soldier whose disappearance was out of character. After her family could not reach her, they contacted her leadership at Fort Hood, alleged to have responded to the family's concerns apathetically. Vanessa's family eventually secured a lawyer, generated media attention, and approached congressional leaders in Texas to help find her. During the investigation, Guillén's family claimed that fellow soldiers were sexually harassing her, but she was afraid to report the abuse out of fear of not being believed. Two months after her disappearance, Guillén's body was found in a shallow grave by the Leon River in Texas. Investigators soon discovered that a fellow soldier, Specialist Aaron Robinson, brutally murdered Guillén on post and transported her body to the Leon River, where he and his girlfriend dismembered and disposed of her body. Guillén's family claimed that Robinson was one of the men sexually harassing her and believed that Guillén was going to report him. Unfortunately, Robinson's motive will remain unknown as he escaped police custody and was killed during a subsequent altercation.[22]

In August of 2020, Sergeant Elder Fernandes died by suicide at Fort Hood after reporting a sexual assault committed by a superior. Fernandes reported a superior for inappropriately touching him and was subsequently transferred to another unit. Peers reported Fernandes to be suicidal after being hazed and bullied for reporting the assault. He was found dead hanging from a tree in Temple, Texas, shortly after seeking behavioral health for the mental distress caused by being assaulted.[23]

The cancer of sexual harassment and assault in the Army metastasized in 2020, igniting a military social justice movement. The deaths and alleged sexual abuses of both Fernandes and Guillén drew outrage within the military and veteran community, igniting a #MeToo social media movement. Hundreds of thousands of victims flocked to social media to share stories of sexual abuses endured in the military, reflecting a culture tolerant of sexual harassment and assault.[24] The national outrage sparked by

---

[20] Appendix F: Sexual Harassment, in Department of Defense Annual Report on Sexual Assault in the Military Fiscal Year 2019 (Washington, DC: U.S. Department of Defense, 2020), 3–6.
[21] Bastian, Lancaster, and Reyst, Department of Defense 1995 Sexual Harassment Survey, iii.
[22] Jennifer Steinbauer, "A #MeToo Moment Emerges for Military Women After Soldier's Killing," New York Times, 30 July 2020.
[23] Rachel Treisman, "Body of Missing Fort Hood Soldier Elder Fernandes Found a Week After Disappearance," National Public Radio, 26 August 2020.
[24] Steinbauer, "#MeToo Moment Emerges for Military Women After Soldier's Killing."

the deaths of Guillén, Fernandes, and other soldiers at Fort Hood prompted Congress to direct the Fort Hood Independent Review Committee to examine the culture and climate that lead to these tragic incidents.

In the words of Ryan McCarthy, former secretary of the Army, "The murder of Specialist Vanessa Guillen shocked our conscience and brought attention to deeper problems within the culture of the US Army."[25] In the wake of 30 years of sexual misconduct scandals, 2020 has clearly demonstrated that the U.S. Army can no longer forgo an in-depth examination of organizational culture.

Problem Statement

The Army has more resources at its disposal than ever to combat sexual harassment and assault, yet statistics continue to increase, and abuses remain largely unreported. Sexual harassment and assault misalign with the Army's formal culture. Sexual harassment and assault violate every Army value and are a punishable offense within the Uniformed Code of Military Justice (UCMJ). The time and resources devoted to the SHARP Program across the Army are visible evidence of the Army's dedication to prevent sexual harassment and assault. However, over the past 30 years, sexual harassment and assault scandals in the Army continue to make national news headlines while leaders blame the incidents on individual behavior.

With a narrowed focus on individual behavior, Army leaders fail to examine the aspects of Army culture that enable the behavior in the first place. For instance, in 2019, nearly 4,000 men committed a confirmed act of sexual assault within DOD, which does not even account for unreported incidents.[26] Sexual harassment and assault in the U.S. Army is a men's issue more than a women's issue. However, the role of men and masculine attitudes and belief systems within the Army often escape in-depth scrutiny.

To overcome an endemic of sexual harassment and assault, the U.S. Army must understand the aspects of its culture that have enabled behaviors contributing to an engrained pattern of sexual harassment and assault within the organization. Attitudes and belief systems drive organizational behavior. Thus, the relationship between masculine attitudes and belief systems and sexual harassment and assault within the Army must be explored. Programs, policies, procedures, and resources allocated to the SHARP Program will continue to be a band-aid solution to the Army's sexual harassment and assault endemic until organizational culture change occurs.

Purpose and Scope of Study

The purpose of this study is to understand how hegemonic masculinities embedded in Army culture influence the occurrence of sexual harassment and assault. This study

---

[25] Ryan D. McCarthy, "DOD Briefing on Findings and Recommendations of the Fort Hood Independent Review Committee," AmericanRhetoric.com, 8 December 2020, hereafter McCarthy briefing.
[26] Appendix D: *Aggregate DOD Data Matrices*, in *Department of Defense Annual Report on Sexual Assault in the Military Fiscal Year 2019*, 6.

seeks to understand the relationship between hegemonic masculinities and patterns of sexual harassment and assault in U.S. Army organizational culture through a case study analysis of the *Report of the Fort Hood Independent Review Committee*. Additionally, this study seeks to understand what aspects of Army culture and climate prevent Army leaders from seeing signs and symptoms of sexual harassment and assault, such as hostile work environments to women. Finally, it recommends strategies to help Army leaders eliminate behaviors that contribute to sexual harassment and assault and erode trust within the organization. It does not provide a quantitative assessment of sexual harassment and assault data in the U.S. Army, nor does this study analyze current sexual harassment and assault prevention programs in the U.S. Army.

The most significant limitations of this study are time and resources. It seeks to understand the influence of hegemonic masculinities on patterns of sexual harassment and assault observed within the climate and culture of Fort Hood. Different installations within the Army may have variances in how culture and climate contribute to sexual harassment and assault. Due to time and resource factors, this study focuses solely on the culture and climate of Fort Hood to understand the role of hegemonic masculinities in perpetuating culture and climate that tolerate sexual harassment and assault. Further research is required to validate if this culture and climate are systemic across the Army. This study does not include new interview or survey data. The research timeline only allowed for a case study analysis of the independent review of Fort Hood's command climate and culture.

Sexual harassment and assault within the U.S. Army is an abhorrent violation of all that the Army values, yet its existence continues systemically within the organization. The military #MeToo movement has taken root within the public, military and veteran communities, Congress, and DOD, demanding accountability and justice for victims and an end to sexual abuses endured as a price for military service.[27] The sexual harassment and assault endemic within the Army is a great risk to losing public trust and confidence. This study intends to provide Army leaders with an analysis of organizational culture, climate, social norms, and behaviors that contribute to sexual harassment and assault. These insights can potentially inform program and policy design and provide a platform to facilitate the organizational culture change necessary to eliminate sexual harassment and assault and restore trust within the profession.

Definitions of Terms

Artifacts: Visible products of a group, such as architecture, language, technology, style, clothing, manners of address, myths, stories, published lists of values, and observable rituals and ceremonies.[28]

Climate: The feeling that is conveyed in a group by the physical layout and the way in which members of the organization interact with each other, with customers, or

---

[27] Steinbauer, "#MeToo Moment Emerges for Military Women After Soldier's Killing."
[28] Edgar H. Schein, *Organizational Culture and Leadership*, 5th ed. (Hoboken, NJ: Wiley, 2017), 17.

with other outsiders. Climate is sometimes included as an artifact of culture and is sometimes kept as a separate phenomenon to be analyzed.[29]

Culture: A pattern of shared basic assumptions learned by a group as it solved its problems of external adaptation and internal integration, which has worked well enough to be considered valid, and therefore, to be taught to new members as the correct way to perceive, think, and feel in relation to those problems.[30]

Espoused beliefs or values: Ideals, goals, values, aspirations, ideologies, and rationalizations.[31]

Hegemonic masculinities: Hegemonic masculinity is a concept that originated in the 1980s to highlight the existence of social norms and cultural rituals that promote a favorable social condition of men over women. Further, the concept presents the idea that all men position themselves culturally to benefit from these favorable social conditions by subjugating themselves to behavior codes that allow social dominance to continue, even if it is to others' detriment. From an ideological perspective, hegemonic masculinity is a version of manhood constructed on the idea that to be a "real man," one must be dominating, heterosexual, display violent and aggressive behavior, and restrain outward displays of vulnerable emotions such as crying. Additionally, hegemonic masculinity requires men to exhibit strength and toughness and be competitive and successful.[32]

Sexual assault: Intentional sexual contact characterized by the use of force, threats, intimidation, or abuse of authority or when the victim does not or cannot consent. The term includes a broad category of sexual offenses consisting of the following specific UCMJ offenses: rape, sexual assault, aggravated sexual contact, abusive sexual contact, forcible sodomy (forced oral or anal sex), or attempts to commit these offenses.[33]

Sexual harassment: Conduct that involves unwelcome sexual advances, requests for sexual favors, and deliberate or repeated offensive comments or gestures of a sexual nature that includes:

- submission to such conduct is made either explicitly or implicitly a term or condition of a person's job, pay, or career;
- submission to or rejection of such conduct by a person is used as a basis for career or employment decisions affecting that person; or
- such conduct has the purpose or effect of unreasonably interfering with an individual's work performance or creates

---

[29] Schein, *Organizational Culture and Leadership*, 17.
[30] Schein, *Organizational Culture and Leadership*, 17.
[31] Schein, *Organizational Culture and Leadership*, 17.
[32] R. W. Connell and James Messerschmidt, "Hegemonic Masculinity: Rethinking the Concept," *Gender and Society* 19, no. 6 (December 2005): 832, https://doi.org/10.1177/0891243205278639.
[33] *Army Command Policy*, Army Regulation 600-20, (Washington, DC: U.S. Department of the Army, 2020).

an intimidating, hostile, or offensive working environment; and
- is so severe or pervasive that a reasonable person would perceive, and the victim does perceive, the environment as hostile or offensive.
- any use or condonation, by any person in a supervisory or command position, of any form of sexual behavior to control, influence, or affect the career, pay, or job of a member of the armed forces or a civilian employee of the DOD.
- any deliberate or repeated unwelcome verbal comment or gesture of a sexual nature by any member of the armed forces or civilian employee of the DOD.[34]

Underlying assumptions: Unconscious, taken-for-granted beliefs, and values.[35]

---

[34] *Army Command Policy*, AR 600-20, 7-7.
[35] Schein, *Organizational Culture and Leadership*, 17.

Chapter Two
Literature Review

In order to organize the literature, this chapter is broken down into three major areas: formal and informal levels of organizational culture, the relationship between organizational culture and sexual harassment and assault, and the relationship between hegemonic masculinities and sexual harassment and assault.

Formal and Informal Levels of Organizational Culture

To understand the relationship between organizational culture and sexual harassment and assault, examining what constitutes organizational culture is warranted. The author selected organizational culture models taught by the U.S. Army Command and General Staff College due to this study's short time frame, and the thesis uses Edward Schein's and Linda Treviño and Katherine Nelson's models of organizational culture to define formal and informal levels of Army culture.

Schein is a world-renowned social psychologist whose work implores researchers, leaders, academics, and other readers to understand that the concept of culture leads us to see patterns in social behavior.[36] Because this thesis focuses on unraveling the pattern of sexual harassment and assault in the Army, Schein's book *Organizational Culture and Leadership* serves as the framework to define its formal culture.

Schein describes culture in terms of three levels: artifacts, espoused beliefs and values, and basic underlying assumptions. The levels of culture vary in their degrees of visibility to the observer.[37]

Artifacts are described as the visible, feelable level of culture. Artifacts can include language, values statements, emotional displays, rituals, and ceremonies. Organizational climate resides in the artifact level of culture. Climate is the feeling conveyed in a group by the physical layout and how members of the organization interact with each other, with customers, or with other outsiders. Artifacts are the most observable aspects of culture but are difficult to decipher. The meanings of artifacts only become clear when people explain why things are done a certain way, which will uncover the next level of culture.[38]

Espoused beliefs and values are what drive how a group or organization accomplishes its core tasks. Espoused beliefs and values can range from rationalizations on how to solve problems and operate to organizational value statements and behavior standards.[39] For example, the Army's "Soldier's Creed" expresses the warrior ethos "I will never quit," thus creating an espoused belief that quitting is unacceptable under any circumstance.

---

[36] Schein, *Organizational Culture and Leadership*, xiii.
[37] Schein, *Organizational Culture and Leadership*, 17.
[38] Schein, *Organizational Culture and Leadership*, 3, 17.
[39] Schein, *Organizational Culture and Leadership*, 19.

Basic underlying assumptions are the deepest, most unconscious level of culture that ultimately determine organizational and individual behavior, perceptions, thoughts, and feelings. Basic assumptions are solutions to problems that have become so ingrained that alternative solutions are inconceivable.[40] Another way to describe basic assumptions is that they are implicit, often unexpressed, ideas that are unconsciously and uncritically taken as true and factual that guide behavior by telling group members how to perceive, think about, and feel about things. Basic assumptions tend to be nondebatable until radical evidence proves a more effective solution to a problem.[41] For example, it would be inconceivable for a couple to have a child before marriage in a religious society. The basic assumption is that marriage must occur to start a family. This assumption is so deeply engrained that most religious societies consider having a child before marriage a sin. This might appear to be a dated example, as many people in present times choose to have children before marriage, however, in many religions, such as Catholicism, it is considered a sin to have children before marriage. This example illustrates the psychology of basic assumptions and why culture has so much power over behavior.

Reexamining basic assumptions is an anxiety-inducing process, so people tend to perceive the world in cohort with basic assumptions, even if it means denying reality. Once culture prescribes a set of basic assumptions in terms of what to pay attention to, what things mean, how to react emotionally to the world, and what actions to take in certain situations, a mental model or lens of how people view the world is formed. Mental models and lenses tend to become ingrained because individuals and groups are most comfortable with others who share similar views.[42] This highlights Schein's key insight that culture's power lies in the fact that assumptions are shared and mutually reinforced. This means that in most instances, it usually takes a third party with experience in different cultures to illuminate underlying basic assumptions within an organization.[43]

Of important note, all basic assumptions do not necessarily remain unchanged. Rationalizations can be disproved using evidence. For example, some people used to think the world was flat. Global culture has since changed the basic assumption to *the earth is round*. Ideals, goals, values, and aspirations cannot be validated or invalidated in the same way and so are much harder to change. Recall the example of religious cultures that deem bearing children before marriage a sin. Even as society has accepted having children before marriage, most religious societies and cultures still consider the choice inconceivable and a sin because of the ideal that a marriage is the foundation of a strong family. This ideal cannot be proved or disproved using evidence the same way that a rationalization can.

---

[40] Schein, *Organizational Culture and Leadership*, 20.
[41] Schein, *Organizational Culture and Leadership*, 22.
[42] Schein, *Organizational Culture and Leadership*, 22.
[43] Schein, *Organizational Culture and Leadership*, 24.

So how do leaders influence and change organizational culture, considering how difficult it is to influence basic, underlying assumptions? Schein's research proposes that leaders must use primary embedding mechanisms to teach their organizations how to perceive, think, feel, and behave based on the conscious and unconscious convictions held by the leader.[44] Within Army organizational culture, a leader's conscious and unconscious convictions are expected to be linked to the Army Values. Drawing on Schein's primary embedding mechanisms, Army leaders integrate the Army Values and tenets of the SHARP Program into organizational culture through what they pay attention to; how they react to crisis and allocate resources; and through deliberate role modeling, teaching, and coaching.[45] Leaders must consistently employ these tools because if their pattern of attention is inconsistent, subordinates will use other signals or their own experience to decide what is important, leading to more diverse assumptions and more subcultures within larger organizational culture.[46] This cursory understanding of primary embedding mechanism helps identify opportunities for Army leaders to use these tools to ensure that organizational values align with behaviors.

Sexual harassment and assault are certainly unethical behaviors, so a discussion of ethics in organizational culture requires inclusion in the literature review. In their book *Managing Business Ethics: Straight Talk about How to Do It Right*, Linda K. Treviño and Katherine A. Nelson present a framework of ethical culture within the context of the broader organizational culture. The main idea they present is that ethical culture is an aspect of organizational culture that represents the way employees think and act in ethics-related situations. Treviño and Nelson propose that ethical culture and decisionmaking are primarily driven by employee socialization: the process of learning the way the organization does things. Employee socialization can occur through various means, such as formal training and mentorship, but also through daily interactions with peers and superiors, which establishes behavioral norms. The broad theory of socialization is that generally people behave in ways consistent with culture because they are expected to.[47] Treviño and Nelson also propose that individual behavior within an organization can also be driven by internalization, where individuals adopt cultural standards as their own.[48]

Socialization and internalization are important in understanding ethical and unethical behavior because employees can be socialized into behaving unethically, especially when they do not have the life experience to know the difference between ethical and unethical behavior.[49] For example, if a young soldier hears everyone

---

[44] Schein, *Organizational Culture and Leadership*, 183.
[45] Schein, *Organizational Culture and Leadership*, 183.
[46] Schein, *Organizational Culture and Leadership*, 188.
[47] Linda K. Treviño and Katherine A. Nelson, *Managing Business Ethics: Straight Talk about How to Do It Right*, 7th ed. (Hoboken, NJ: Wiley, 2017), 158–60.
[48] Treviño and Nelson, *Managing Business Ethics*, 161.
[49] Treviño and Nelson, *Managing Business Ethics*, 161.

Figure 6.1. Multisystem Ethical Culture Framework

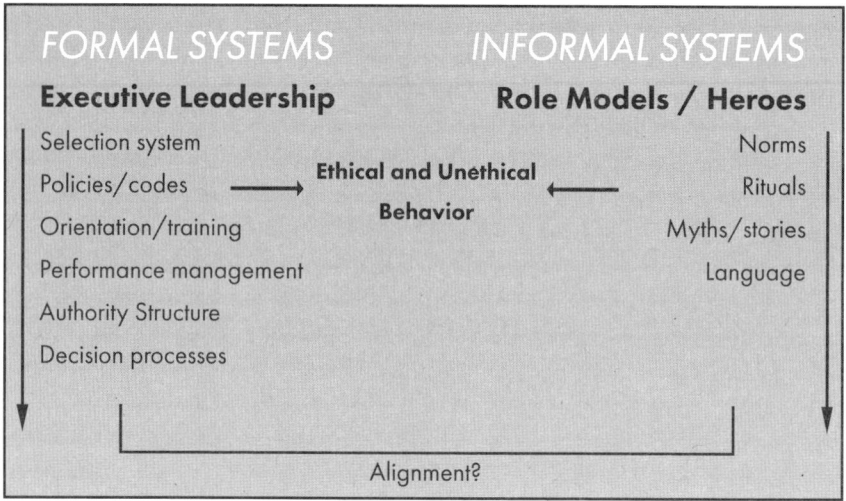

FORMAL SYSTEMS	INFORMAL SYSTEMS
Executive Leadership	**Role Models / Heroes**
Selection system	Norms
Policies/codes → **Ethical and Unethical** ←	Rituals
Orientation/training **Behavior**	Myths/stories
Performance management	Language
Authority Structure	
Decision processes	

Alignment?

Source: Based on Linda K. Treviño and Katherine A. Nelson, *Managing Business Ethics: Straight Talk about How to Do It Right*, 7th ed. (Hoboken, NJ: Wiley, 2017), 161.

around them using profanity in daily communications, they will likely do the same, even if they feel uncomfortable, because if they do not partake, they would likely be ostracized within the group. This thesis uses Treviño and Nelson's Multisystem Ethical Culture Framework to understand how ethical culture is created and sustained.

The Multisystem Ethical Culture Framework (figure 6.1) illustrates that ethical culture is balanced between the interaction of formal and informal organizational culture systems. For organizations to send a clear message of what constitutes ethical culture and behavior, both the formal and informal organizational cultures must be aligned.[50]

Revisiting the previous example of a young soldier being immersed in a unit where all the leaders use profanity, according to the Army's values, excessive use of profanity is certainly disrespectful and contrary to the Army value of respect. Performance evaluations within the U.S. Army require all officers and noncommissioned officers (NCOs) to be evaluated on their compliance with Army values. If all the leaders using profanity and disrespectful language are given a substandard evaluation for their failure to act in accordance with the Army Values by using profanity, then the unit senior leadership would be demonstrating a clear ethical alignment with formal and informal cultural systems. This action would send a message that profanity is not acceptable in accordance with Army values and behavioral expectations. However, if the unit senior leadership does the opposite and gives spectacular evaluations to the leaders using profanity, then an ethical misalignment of culture occurs. This ethi-

---

[50] Treviño and Nelson, *Managing Business Ethics*, 162.

cal misalignment sends the message that leaders can behave in ways that are not in accordance with the Army Values and still receive strong performance evaluations. While the use of profanity may seem like a minor issue, the example illustrates a serious point: leaders create ethically aligned culture by sending formal and informal messages about what behavior is and is not acceptable. Army leaders can undoubtedly benefit from applying the principles of ethically aligned organizational culture to address the current challenges with sexual harassment and assault.

Impact of Culture on Sexual Harassment and Assault

This section describes the relationship between organizational culture and sexual harassment and assault by reviewing themes and patterns of organizational culture and behavior closely linked to sexual harassment and assault. The first article that warrants discussion is Juanita Firestone and Richard Harris's article from the *Armed Forces and Society* journal, "Changes in Patterns of Sexual Harassment in the U.S. Military: A Comparison of the 1988 and 1995 DoD Surveys." While this article was published in 1999, it offers relevant historical insights applicable to current challenges with sexual harassment and assault faced by the Army.

Firestone and Harris's main purpose was to compare and contrast the 1988 and 1995 DOD sexual harassment surveys. While the comparison of surveys showed very little change over a seven-year period, Firestone and Harris offer several keen insights on organizational patterns of behavior. Firestone and Harris argue that sexual harassment will persist until the DOD stops conceptualizing sexual harassment as individual behavior while ignoring organizational norms that tolerate sexual harassment as acceptable behavior.[51]

The article emphasizes that organizational norms within the military have traditionally focused on male bonding rituals designed to build group cohesion, which is a highly valued aspect of military culture. Firestone and Harris then suggest that women and men who do not emulate hypermasculine traits are generally thought to be unaccepting of male bonding rituals, which causes the dominant group to shift focus by finding ways to exclude those groups from being a part of unit cohesion. This basic assumption allows environmental harassment to become a covert method to restrict women and some men's acceptance to the dominant group while also working to undermine credible reports of sexual harassment as "false accusations." Firestone and Harris end the article with a stark warning that to make real progress in decreasing workplace sexual harassment the DOD must work immediately to confront the hypermasculine military culture creating a hostile climate toward women and men who do not conform to those ideals.[52]

---

[51] Juanita M. Firestone and Richard J. Harris, "Changes in Patterns of Sexual Harassment in the U.S. Military: A Comparison of the 1988 and 1995 DoD Surveys," *Armed Forces and Society* 25, no. 4 (Summer 1999): 613, https://doi.org/10.1177/0095327X9902500405.
[52] Firestone and Harris, "Changes in Patterns of Sexual Harassment in the U.S. Military," 617, 625–27.

Interestingly enough, a similar theme of a hypermasculine culture and organizational norms that exclude women surfaced in Stephanie Switzer's doctorate thesis "Sexual Harassment and Sexual Assault in the Military." In it, Switzer highlights several themes that influence sexual harassment and assault in the military:

1. A masculinized culture where gender hostility is pervasive and sexually aggressive behaviors are tolerated;
2. Men outnumbering women;
3. Unit cohesion that protects perpetrators and punishes women for reporting through various forms of retaliation and blaming;
4. The abuse of rank and power to perpetrate abuses or ignore abuses completely; and
5. Organizational climate that takes a laissez-faire approach to responding to formal and informal reports of sexual harassment and assault.[53]

Switzer's dissertation was written to help military leaders recognize and modify personal biases and beliefs that contribute to an organizational culture that sustains high rates of sexual harassment and assault. Her work is incredibly relevant to this study in identifying similar themes within the U.S. Army.

Another 2007 study titled "Attitudes Toward Women and Tolerance for Sexual Harassment Among Reservists" revealed that attitudes and beliefs about women's abilities and the overall acceptance level of women serving in the military were independently related to tolerance for sexual harassment and assault.[54] Essentially, if those surveyed conveyed the attitude that women could and should serve in the military, they were much less tolerant of sexual harassment and assault than people who conveyed the attitude that women did not possess the capabilities to serve and should not do so. The study's findings, which were drawn through original survey data conducted among veterans of the U.S. military, continue to draw on previous themes that gender hostilities within organizational culture promote an environment tolerant of sexual harassment and assault.

The most recent study available that addresses sexual harassment and assault in the context of organizational culture is a 2017 Rand study, *Improving Oversight and Coordination of Department of Defense Programs that Address Problematic Behaviors Among Military Personnel*, to assist the DOD with developing a framework to prevent and modify six problematic behaviors: sexual harassment, sexual assault, unlawful discrimination, substance abuse, suicide, and hazing.[55] The report conducted a

---

[53] Stephanie Lise Switzer, "Sexual Harassment and Sexual Assault in the Military" (PhD diss., University of Hartford, 2007), 28–29.
[54] Dawne Vogt, Tamara A. Bruce, Amy E. Street, and Jane Stafford, "Attitudes Toward Women and Tolerance for Sexual Harassment Among Reservists," *Violence Against Women* 13, no. 9 (September 2007): 879–900, https://doi.org/10.1177/1077801207305217.
[55] Jefferson P. Marquis et al., *Improving Oversight and Coordination of Department of Defense Programs that Address Problematic Behaviors among Military Personnel: Final Report* (Santa Monica, CA: Rand, 2017), ix.

behavioral analysis presenting significant empirical evidence that attitudes seem to predict problematic behavior best when organizational culture also supports the behavior. In other words, someone is more likely to engage in problematic behavior, such as sexual harassment, if that person perceives that peers and leaders explicitly or implicitly condone those actions. Conversely, people who might be initially inclined toward problematic behavior can be dissuaded if the organizational climate is clearly opposed to such behavior.[56] The report also notes that few academic studies exist examining the relationships among problematic behaviors, establishing a clear need for this study and others proposing organizational and cultural approaches to improve servicemembers' well-being.

Impact of Hegemonic Masculinities on Sexual Harassment and Assault

Next, this study wants to understand previously identified themes of sexual aggression, masculinized culture, and gender hostility to further characterize sexual harassment and assault within the context of Army organizational culture. Foundational knowledge of gender order theory is required to understand the root cause of sexual aggression, abuse, and violence. Recall in the previous section that a theme of hypermasculine culture was identified multiple times as a contributing factor to organizational sexual harassment and assault. So, what exactly is *masculine culture*, and what is the role of masculinities in sexual harassment and assault?

In a joint article titled "Hegemonic Masculinity: Re-Thinking the Concept," R. W. Connell and James Messerschmidt present the concept of hegemonic masculinity and discuss its impact and evolution within social science and gender study research over the past 30-plus years. Connell's research on masculinities and social power relations is the most widely accepted framework within sociological gender theory studies. For this reason, this thesis utilizes Connell's definitions of hegemonic masculinity and gender theory concepts as a theoretical framework to define and understand hegemonic masculinities within Army culture. The concept of hegemonic masculinity originated from research conducted by Connell throughout the 1980s focused on understanding social inequality. Connell's research proved through empirical evidence that within all local cultures, a normative, dominant ideal of what it meant to be a man (masculinity) and a woman (femininity) existed. Further, Connell's study identified the existence of gender hierarchies within a culture. For example, multiple forms of masculinity may exist within a culture, but one form of masculinity always serves as the dominant or hegemonic masculinity, and those that embodied hegemonic masculinity within a culture were the dominant group (figure 6.2).[57]

---

[56] Marquis et al., *Improving Oversight and Coordination of Department of Defense Programs that Address Problematic Behaviors among Military Personnel*, xi.
[57] Connell and Messerschmidt, "Hegemonic Masculinity," 830.

Figure 6.2. R. W. Connell's Hierarchy of Masculinities

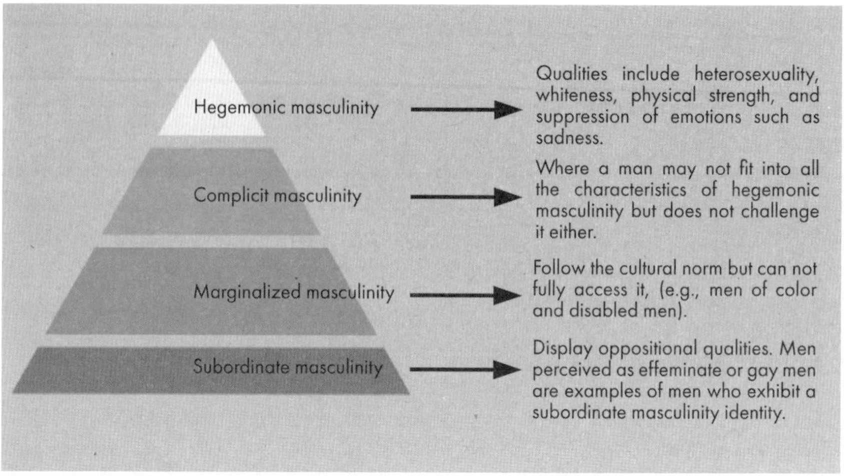

Hegemonic masculinity	Qualities include heterosexuality, whiteness, physical strength, and suppression of emotions such as sadness.
Complicit masculinity	Where a man may not fit into all the characteristics of hegemonic masculinity but does not challenge it either.
Marginalized masculinity	Follow the cultural norm but can not fully access it, (e.g., men of color and disabled men).
Subordinate masculinity	Display oppositional qualities. Men perceived as effeminate or gay men are examples of men who exhibit a subordinate masculinity identity.

Source: based on Raewyn Connell, "Masculinities," Raewyn Connell (personal website).

Connell and Messerschmidt define hegemonic masculinity as a distinguished form of masculinity that embodies the current, most honored way of being a man, and consciously or unconsciously, all other men position themselves to benefit from the social gains of hegemonic masculinity. Hegemonic masculinity also has a role within sociological power structures such as the military, governments, and private corporations. In political sociology, hegemonic masculinity is widely accepted as the pattern and practices within organizational culture (how things are done, i.e., artifacts and norms), which allows men's dominance over women and subordinate masculinities to continue.[58]

It is important to note that while only a small portion of men might enact hegemonic masculinity, a dominant form of masculinity is engrained and normative within every culture and often results in the construct of a patriarchal gender system. All men receive the benefits of patriarchy even when enacting subordinate masculinities, such as complicit, marginalized, and subordinate masculinities. Connell and Messerschmidt highlight that the subordination of alternate masculinities paired with compliance among heterosexual women is what makes the concept of hegemony so powerful. While hegemony is not synonymous with violence, it can be supported through force, but most importantly, enacting and/or compliance with hegemonic masculinity is required to ascend to the top of social and power structures within cultures and institutions.[59]

Connell and Messerschmidt note that the harm of hegemonic masculinities is patterns of aggression and abuse enacted by individuals and groups to pursue dom-

---

[58] Connell and Messerschmidt, "Hegemonic Masculinity," 832.
[59] Connell and Messerschmidt, "Hegemonic Masculinity," 832.

inance, power, and social ascendency. The struggle for hegemony, not hegemonic masculinity itself, links hegemonic masculinities to violence and aggression. Further, the normalization of violent, abusive, and other dehumanistic and aggressive behavior of men and boys within cultures and institutions is the primary driver of highly visible social mechanisms such as oppressive policies, behaviors, and widely accepted practices directed at subordinate groups such as gay men, minorities, and women. Examples can range from the dismissal of school-age boys' aggression through the "boys will be boys" mentality to criminalizing homosexual conduct.[60]

More importantly, Connell and Messerschmidt argue the less visible mechanisms of hegemony often remove dominant forms of masculinity from the possibility of scrutiny.[61] A major example of this is deeming domestic and sexual violence a women's issue. Globally, women and men are harmed predominantly by men. Understanding what aspects of culture cause abusive and violent behavior in men is imperative to prevent violence. The role of men in and masculinities must be examined and restructured for change to occur. Still, stakeholders within societal institutions largely allow the ideals of hegemonic masculinities to go unexamined and unchecked, perpetuating vicious cycles of abuses, violence, and other counterproductive social ideologies such as gender discrimination, racism, and sexism.

So, what does hegemonic masculinity look like in the military? The article titled "Real Men: Countering a Century of Military Masculinity" by Joshua Isbell discusses the history of idealized masculinities in the context of military service and points out how the U.S. military, in particular, is struggling with discrimination and harassment in the ranks because of an idealized version of what it means to be a "real man." Isbell traces the roots of hegemonic masculinities to Europe before World War I, reminding readers of the invocations of masculine pride that compelled the people of Europe to enter into the war.[62]

Isbell reminds readers that 100 years ago, the nations of Europe challenged young men to prove their manliness, patriotism, and citizenship through military service. "Real men" achieved their status in society by fighting the nation's wars, thus interweaving idealized masculinities and social status with military service. Isbell argues that this ideology created an unattainable version of successful manhood, creating a tension between the men struggling to achieve idealized masculinity within society and the contributions of other groups such as women, minorities, and men conscripted into service or fulfilling combat support roles in military service.[63]

While striving to achieve the self and societal ideal of masculinity, men serving in combat roles minimize the contributions of women, minorities, and those serving in combat support roles. This climate reinforces unhealthy social norms within the mili-

---

[60] Connell and Messerschmidt, "Hegemonic Masculinity," 834.
[61] Connell and Messerschmidt, "Hegemonic Masculinity," 835.
[62] Joshua Isbell, "Real Men: Countering a Century of Military Masculinity," War Room, 1 March 2019.
[63] Isbell, "Real Men."

tary that the "real men" fight and serve in direct combat roles and other contributions and roles do not matter in the same way. Isbell illustrates a few examples of inequalities driven by hegemonic masculinities, first citing the pervasive use of the term "position other than grunt" (POG) to describe the service and contributions of those not in direct combat roles. The use of this term is meant to reinforce a power dynamic that the service that matters is the service of men fulfilling direct combat roles. Being called a POG is not a term of endearment; its use intends to undermine the contributions of other servicemembers who do not equate to the social definition of masculinity.[64]

Additionally, Isbell attributes the perceived lack of deference from society to the status and'manliness achieved through military service to reoccurring outbursts of aggression and violence from men against women and minorities in both the military and society. Isbell believes many men, both military and nonmilitary, dissatisfied by this lack of societal deference are joining white nationalist organizations seeking to impose regressive race and gender hierarchies to validate their place in society. Isbell cites this example to implore readers to understand the danger of the ideals of hegemonic masculinities. Not only are the ideals unattainable, but the ideals of hegemonic masculinity fail to obtain the inclusiveness the military and society require to maximize performance. Isbell closes his piece by imploring leaders to stop making appeals based on unattainable masculine ideals but instead focus on the fact that military success has always relied on both men's and women's best contributions.[65]

To further understand the role of hegemonic masculinity in military culture, we next turn to *The Organizational Construction of Hegemonic Masculinity: The Case of the US Navy* by Frank Barrett, which discusses the social construction of masculinities within the U.S. Navy. While the focus of this study is the culture within the Army, Barret's article outlines the inner workings of hegemonic masculinity within a militarized culture in great detail, which makes the concepts worthwhile to explore.

Through life history interviews with 27 naval officers who served in surface warfare, aviation, and supply, Barrett identifies that all groups of officers construct definitions of masculinity by highlighting the masculine characteristics necessary and unique to one's career path and why those characteristics are more valuable than others. For example, the naval aviation officers identified themes of autonomy and risk-taking as masculine traits necessary to thrive as a Navy man, while the supply officers identified themes of technical rationality as the most important.

Most interestingly, masculinity was constructed and "proven" through social accomplishment and achieved meaning by drawing a stark contrast to femininity. Essentially, all of the masculinities identified within the naval officer corps achieved meaning in contrast with definitions of femininity.[66] Across all men's interviews, wom-

---

[64] Isbell, "Real Men."
[65] Isbell, "Real Men."
[66] Frank J. Barrett, "The Organizational Construction of Hegemonic Masculinity: The Case of the US Navy," *Gender, Work and Organization* 3, no. 3 (July 1996): 129, https://doi.org/10.1111/j.1468-0432.1996.tb00054.x.

en are depicted as emotionally unstable, less physically capable, and unable to handle harsh living conditions, which is consistent with masculine socialization in Western culture. Essentially boys are taught from a very young age that being a man has no other definition than not being a woman—masculinity is defined more by what one is not rather than what one is.[67]

Barrett proposes that within military culture, ritualistic displays of hegemonic masculinity often become a way to exclude women from social activities or "othering" women and normalize degrading behavior and language directed at women. The construct of masculinity essentially becomes an invisible, unconscious strategy that undermines women's abilities to meaningfully contribute to the defense of the United States by depicting them as innately unsuited for military service.[68]

Like Isbell, Barrett argues that appealing to the ideals of hegemonic masculinity is a dangerous game for leaders in the U.S. military. Barrett's study clearly identifies the existence of a competitive masculine culture within the U.S. military in which men and women must continuously demonstrate competence that many men in the military also equate to their status as men. This competitive culture constantly increases the threshold to demonstrate masculinity, which can result in violent or aggressive behavior, especially against others such as women and homosexual men. Competitive masculine culture reinforces dehumanizing language as socially acceptable, setting conditions for a climate tolerant of further abuses. Further, competitive masculine cultures are detrimental not only to women and subordinated masculinities such as minority and gay men but also to the men who feel the social pressures to participate in these masculinity contests. Competitive masculine culture encourages a cycle of continual defensive posturing, validating oneself through outperforming the team and negating the contributions of others.[69] This type of environment is detrimental to teamwork, collaboration, and comradery necessary for the U.S. military to solve difficult and dangerous problems.[70]

Hegemonic masculinity and forced gender roles are ingrained in both men's and women's consciousness from a very young age. Acceptance of sexual aggression and other problematic behaviors such as racism, sexism, hazing, and bullying are direct results of hegemonic masculinities playing out in organizational climates. Further, the trained acceptance of the aggressive and abusive social norms of hegemonic masculinity by society, especially those in leadership, only perpetuates its vicious cycle.

Summary

A multitude of literature exists within professional, academic, and military institutions regarding the relationship between organizational culture and sexual harassment

---

[67] Barrett, "Organizational Construction of Hegemonic Masculinity," 140.
[68] Barrett, "Organizational Construction of Hegemonic Masculinity," 140.
[69] Barrett, "Organizational Construction of Hegemonic Masculinity," 142.
[70] Isbell, "Real Men."

and assault. While several military-specific studies hint at the idea of hegemonic masculinities as a challenge in combatting sexism, sexual harassment and assault, and other diversity and inclusion initiatives, most of the studies do not explore the ideology in great detail. A focused study attempting to understand and explain the relationship between hegemonic masculinities and sexual harassment and assault within U.S. Army organizational culture does not exist, making this project worthwhile.

Chapter Three
Research Methodology

Ultimately, the findings of this study will be critical to determine the relationship between hegemonic masculinities and sexual harassment and assault. This chapter is divided into three sections to describe the research methodology. As outlined earlier, the construct of this study focuses on a case study analysis of the *Report of the Fort Hood Independent Review Committee*, which will serve as a reflection of broader Army culture.

Schein's definitions of organizational culture serve as this study's research framework to identify and create an understanding of the formal culture as it relates to sexual harassment and assault within the Army. Using Schein's levels of culture, the analysis creates an initial understanding of the formal systems and policy within the Army dedicated to preventing sexual harassment and assault. A foundational understanding of Army formal culture lays the groundwork to identify and frame informal and formal culture misalignments observable within the *Report of the Fort Hood Independent Review Committee*.

Next, drawing from Schein's model of organizational culture, this study frames the influence of hegemonic masculinity at the underlying assumption level of formal Army culture. Connell and Messerschmidt's definition of hegemonic masculinity described previously informs this analysis. The link between hegemonic masculinity and the underlying assumption level of culture creates the initial understanding necessary to analyze informal culture in greater detail within the Fort Hood Independent Review Committee's (FHIRC) report. Connell's Hierarchy of Masculinities serves as the analytical lens to identify and interpret any norms and social mechanisms associated with hegemonic masculinities observable within the informal culture of Fort Hood as described by the FHIRC.

Sexual assault and harassment, the associated norms and language, and the social mechanisms of hegemonic masculinity are most visible within informal organizational culture. Treviño and Nelson's Multisystem Ethical Culture Framework serves as an additional analytical lens to describe the relationship between the norms, language, and social mechanisms of hegemonic masculinities and sexual harassment and assault observable within the informal culture described in the FHIRC report. Lastly, Treviño and Nelson's framework helps identify and frame any observable misalignments between formal and informal culture within the Fort Hood case study.

This study uses a multifaceted, integrated research methodology consisting of a descriptive and explanatory case study analysis of the FHIRC report to answer the research questions. Descriptive case study methods serve the primary purpose of describing a phenomenon or case in a real-world context. Subsequently, explanatory case studies focus on explaining how or why some condition came to be.[71]

To answer the primary research question, this thesis first employs the descriptive case study method and analytical lenses to identify patterns of hegemonic masculine ideology within the culture and climate identified at Fort Hood to provide readers with a real-world context of the phenomena. Next, this thesis explains the effects of hegemonic masculine ideologies on observable patterns of sexual harassment and assault at Fort Hood using the explanatory case study methodology. It then uses the same methodology and analytical lenses to address the secondary research question to identify cultural themes related to sexual harassment and assault that prevent leaders from identifying the signs and symptoms of sexual assault and harassment. This analysis aims to bring deeper cultural issues into the forefront of Army leaders' shared consciousness and provide a cursory explanation of the phenomena. Further, this analysis provides observations and identifies cultural themes within the FHIRC report to inform thematic analysis for future qualitative research studies.

This study's final research question intends to provide Army leaders with viable recommendations to mitigate and eliminate dangerous organizational behavior that contributes to sexual harassment and assault and undermines trust. Identifying how cultural ideologies and biases affect Army leaders' ability to see dangerous problems such as sexual harassment and assault sets the stage to propose recommendations they can act on immediately to eliminate organizational behaviors that undermine trust and contribute to sexual abuses. Additionally, this analysis provides Army leaders an opportunity to identify other installations struggling with systemic sexual harassment and assault and lack of trust in leadership.

Data Analysis

To address the research questions, the case study analysis of the FHIRC report focuses on the report findings that address the keywords of *command climate, climate, culture, gender, male, female, trust, confidence,* and *sexual assault and harassment*. This coding plan narrows the analysis to the following report findings:

> Finding #1: The implementation of the SHARP Program at Fort Hood has been ineffective due to a command climate that failed to instill SHARP Program core values below the brigade level.
> Finding #2: There is strong evidence that incidents of sexual assault and sexual harassment at Fort Hood are significantly underreported.
> Finding #3: The Army SHARP Program is structurally flawed.
> Finding #5: The mechanics of the Army's adjudication processes involving sexual assault and sexual harassment degrade confidence in the SHARP Program.

---

[71] Robert K. Yin, *Case Study Research and Applications: Design and Methods*, 6th ed. (Los Angeles: Sage, 2018), 45.

Finding #8: The criminal environment within surrounding cities and counties is commensurate with or lower than similar sized areas; however, there are unaddressed crime problems on Fort Hood, because the installation is in a fully reactive posture.

Finding #9: The command climate at Fort Hood has been permissive of sexual harassment/sexual assault.

After analyzing the FHIRC's findings, this study organizes and presents major observations applicable to each of the research questions posed in chapter 1.

Research Feasibility

As previously discussed, this study employs a descriptive case study analysis of Fort Hood's command climate and culture to understand the role of hegemonic masculinities in perpetuating a culture and climate tolerant of sexual harassment and assault as a representation of the broader Army culture. Different installations within the Army may have variances in how culture and climate contribute to sexual harassment and assault. Further research will be required to validate if the aspects of culture and climate contributing to sexual harassment and assault at Fort Hood are systemic across the Army.

Selection of Research Material

The events at Fort Hood in 2020, specifically the alleged sexual harassment and brutal murder of Specialist Vanessa Guillén, prompted a #MeToo movement within the U.S. military, capturing global news headlines.[72] Congressional leadership quickly directed an independent, congressionally mandated investigation into the culture and climate of Fort Hood. The investigators published the *Report of the Fort Hood Independent Review Committee*, which lists the culture and command climate observations and proposed recommendations that serve as the case study for this thesis. This report contains the most relevant and current data on sexual harassment and assault within U.S. Army culture and climate. The severity of the situation at Fort Hood and across the Army concerning systemic sexism and racism most certainly warrants a more in-depth examination and explanation of the report's observations on culture and climate.

The proposed research methodology, feasibility, and selection of research material are broad enough to enable holistic research while also focused enough to answer this study's research questions. At a minimum, the methodology identifies cultural themes and patterns contributing to sexual harassment and assault at Fort Hood to inform thematic analysis for future qualitative studies of U.S. Army culture.

---

[72] Steinbauer, "#MeToo Moment Emerges for Military Women After Soldier's Killing."

Chapter Four
Analysis

A review of the concepts of culture and hegemonic masculinity is warranted to set the context for the case study analysis. Schein's model of organizational culture contains three levels, and the deepest level is a culture's basic underlying assumptions that guide behavior by telling group members how to perceive, think, and feel.[73] These underlying assumptions are implicit norms that drive how a group or organization accomplishes core tasks ranging from how to solve problems and operate to organizational value statements and behavior standards. The specifics of the underlying assumptions are hard to identify, but the espoused beliefs and artifacts of a culture provide indications of those underlying assumptions.[74] Schein describes artifacts as the visible, feelable level of culture, which can include language, values statements, rituals, and observable behaviors within a culture.[75]

Nelson and Treviño add to Schein's definition of culture by categorizing culture as both formal and informal. The public statements and ceremonies of the Army are visible elements of formal Army culture. Nelson and Treviño also describe an aspect of culture that is less defined, less codified as the informal culture. This informal culture is what members of the unit do or say or believe, not because of a policy or formal order but because that's what everybody who wants to be part of the group is doing. An example in the Army are the traditions of the companies, platoons, and squads that are unique or specific to that unit. Most of those traditions and behaviors align with the espoused values of formal Army culture, but some may not, as in the case of hazing.

Hazing is one indication of the influence of hegemonic masculinity on Army culture. Connell and Messerschmidt define hegemonic masculinity as a distinguished form of masculinity that embodies the current, most honored way of being a man. Consciously or unconsciously, men and women position themselves to benefit from the social gains of hegemonic masculinity. From an ideological perspective, hegemonic masculinity is a version of manhood constructed on the idea that to be a "real man," one must be dominating, heterosexual, display violent and aggressive behavior, and restrain outward displays of vulnerable emotions such as crying. Additionally, hegemonic masculinity requires men to exhibit strength and toughness to be competitive and successful.[76] Connell's hierarchy of masculinity identifies that multiple versions of masculinity exist within a culture, but only one form of masculinity is normative and the hegemonic masculinity.[77] Further, those who embodied some or all of the characteristics of hegemonic masculinity within a culture were the dominant group.[78]

---

[73] Schein, *Organizational Culture and Leadership*, 21–22.
[74] Schein, *Organizational Culture and Leadership*, 19.
[75] Schein, *Organizational Culture and Leadership*, 17.
[76] Connell and Messerschmidt, "Hegemonic Masculinity," 832.
[77] Raewyn Connell, "Masculinities," Raewyn Connell (personal website), accessed 13 March 2021.
[78] Connell and Messerschmidt, "Hegemonic Masculinity," 830.

As previously stated, Connell's extensive research in gender order theory states that hegemonic masculinity is normative. All cultures exhibit one if not multiple forms of hegemonic masculinities. Within sociological power structures such as the military, hegemonic masculinity is widely accepted as the pattern and practices within organizational culture, allowing men's dominance over women and men who embody alternate forms of masculinity to continue.[79] Military culture, specifically U.S. Army culture, is not immune to the effects of hegemonic masculinity within both formal and informal culture.

As described in chapter 2, Isbell traces the roots of hegemonic masculinities within the context of military service back to World War I, where government leaders in both Europe and the United States invoked masculine pride of society by challenging men to prove their manliness, patriotism, and citizenship through military service. Real men achieved their status in society by fighting the nation's wars, thus interweaving idealized masculinities and social status with military service.[80] This connection created popular images of masculinity within larger society displaying the soldier as the embodiment of male sex role behaviors.[81] These associations influence larger society but also still exist within the organizational culture of the Army.

For example, General Douglas MacArthur was a domineering, aggressive, and authoritarian officer who emerged as a societal and organizational hero following his accomplishments in the Pacific theater during World War II. Following World War II, MacArthur was relieved of command by President Harry S. Truman in large part because of blatant insubordination and his aggressive approach to expelling the North Korean People's Army (NKPA) and People's Liberation Army of China from South Korea during the Korean War of 1950. MacArthur continually made contradictory statements to the press that undermined President Truman's authority and ultimately thwarted the president's attempt to negotiate a ceasefire when the general ordered his troops to invade North Korea and push the NKPA up past the 38th parallel.[82] MacArthur's actions yielded costly results with almost 1,500 casualties incurred at the hands of the Chinese and the total destruction of Lieutenant Colonel Don C. Faith Jr.'s task force.[83]

Despite all this, MacArthur is still glorified within Army culture today. The General Douglas MacArthur Leadership Award recognizes company-grade officers who demonstrate the ideals for which MacArthur stood: duty, honor, and country. This award is a longstanding Army ritual led annually by the Chief of Staff of the Army. The MacArthur Leadership award is an artifact that symbolizes the Army's conscious and

---

[79] Connell and Messerschmidt, "Hegemonic Masculinity," 832.
[80] Isbell, "Real Men."
[81] Barrett, "The Organizational Construction of Hegemonic Masculinity," 129.
[82] "The Firing of Macarthur," Harry S. Truman Presidential Library, accessed 17 April 2021.
[83] Robert J. Reilly, "Defeat from Victory: Korea 1950," in *The Last 100 Yards: The Crucible of Close Combat in Large-Scale Combat Operations*, ed. Paul E. Berg (Fort Leavenworth, KS: Army University Press, 2019), 138.

unconscious bias that idealizes leaders such as General MacArthur, who embodied hegemonic masculine leadership characteristics despite major flaws in his leadership style that cost significant loss of American life.

Indeed, some of the characteristics of hegemonic masculine leaders like MacArthur, such as aggressiveness and assertiveness, are necessary to achieve success in combat. The harm of hegemonic masculinities and organizational leaders and team members that embody the ideology are patterns of aggression and abuse enacted by these individuals and groups to pursue dominance, power, and social ascendancy.[84] The struggle for hegemony, not hegemonic masculinity itself, links hegemonic masculinities to violence and aggression. Patterns of sexual assault and harassment within organizational culture and climate are equivalent to patterns of aggression and abuse that can result from hegemonic masculinity. The analysis within this chapter identifies visible artifacts, espoused beliefs, and social norms of hegemonic masculinities within the climate and culture of Fort Hood that contributed to patterns of sexual assault and harassment.

The SHARP Continuum of Harm is a visual tool to understand the Army's continuum of acceptable and unacceptable behaviors which may progress to sexual harassment and assault (figure 6.3).

Figure 6.3. U.S. Army SHARP Continuum of Harm

PROFESSIONAL WORK ENVIRONMENT	EARLY WARNING SIGNS	SEXUAL HARASSMENT	SEXUAL ASSAULT
• Engaged leadership • Army Values • Good order and discipline • Dignity and respect • Ethical standards • Accountability • Safe environment • Warrior ethos • Civilian creed **KEEP IT IN THE GREEN!**	• Excessive flirting • Toxic atmosphere • Inappropriate jokes or comments • Disparaging comments on social media • Inappropriate work relationships	• Sending unsolicited naked pictures • Indecent recording or broadcasting • Nonconsensual kissing or touching • Indecent exposure • Indecent viewing • Cat calls • Sexual innuendos • Cornering or blocking • Sexually oriented cadence • Unsolicited sexually explicit texts or emails	• Rape • Abusive sexual contact • Aggravated sexual contact Sexual harassment and sexual assault are both criminal offenses under UCMJ. They reduce a unit's overall mission readiness by destroying trust and unit cohesion.

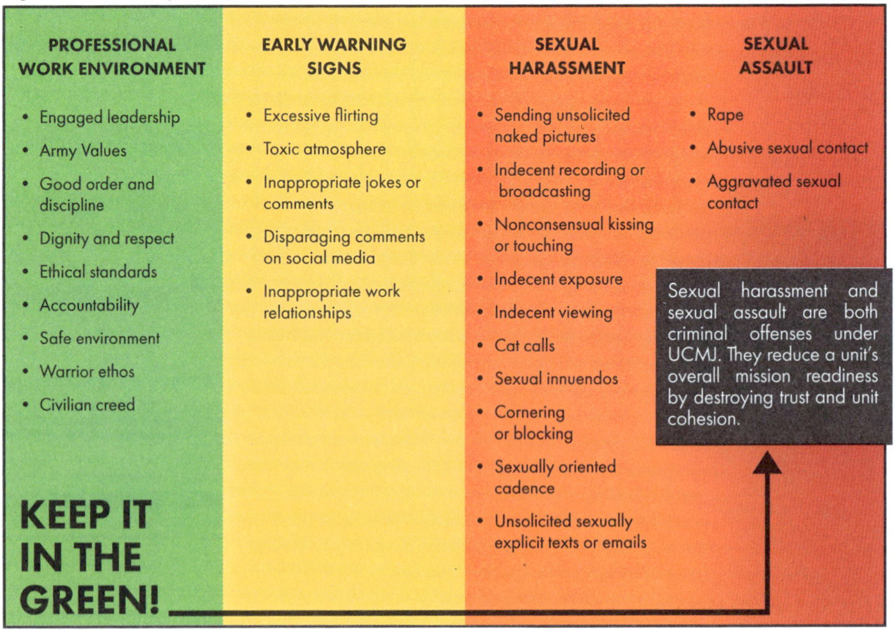

Source: "Army Sexual Harassment/Assault Response and Prevention Continuum of Harm," U.S. Army Resilience Directorate, accessed 17 March 2021.

---

[84] Connell and Messerschmidt, "Hegemonic Masculinity," 834.

As stated at the top of the graphic, sexual harassment and assault reduce a unit's overall mission readiness by destroying trust, teams, and unit cohesion. This graphic and associated SHARP policies are examples of espoused beliefs of formal Army culture. The graphic then depicts and describes the continuum from professional behavior to sexual assault. This Continuum of Harm indicates the Army's acknowledgment that attitudes and behaviors identified as early warning signs can lead to unacceptable behaviors such as sexual assault and other forms of violence. As shown in the graphic, soldiers are expected to "Keep it in the Green" and sustain a professional working environment consistent with all the characteristics described on the left-hand side of the continuum. Leaders must be engaged to sustain a professional working environment and intervene immediately to correct work environments that stray away from professionalism. Additionally, leaders and soldiers are told to report incidents of sexual assault and harassment to SHARP professionals.[85] These espoused beliefs establish clear standards of behavior and provide leaders and soldiers within the U.S. Army guidance on preventing and handling instances of sexual harassment and assault.

By all accounts, Army formal culture is very clear on how soldiers and leaders are expected to think and behave to foster a culture free of sexual harassment and assault. To make things even more apparent, the Army Values and other positive behaviors such as engaged leadership are listed as artifacts that reflect a professional working environment or climate within the SHARP Continuum of Harm.[86] This distinction of what Army formal culture communicates as acceptable culture and climate to prevent sexual assault and harassment serves as a reference throughout the analysis of the FHIRC report to frame misalignments between formal and informal culture.

The SHARP Continuum of Harm is a visible artifact the Army employs to create a foundational understanding that attitudes that allow or enable any forms of harassment are the foundation that can lead to more egregious behaviors such as sexual assault and harassment and other forms of violence.[87] Essentially, basic underlying assumptions at the deepest level of culture ultimately influence observable behaviors such as sexual harassment and assault. The premise of this thesis is to explore why men, in particular, commit sexual harassment and assault within the Army. This warrants a discussion on the relationship between masculinity, which is constructed at the basic assumption level of culture, and its influence on both formal and informal culture.

The primary research question of this thesis is: How do hegemonic masculinities embedded in Army culture affect the occurrence of sexual harassment and assault? This section presents the author's observations drawn from an examination of the cul-

---

[85] "Army Sexual Harassment/assault Response and Prevention Continuum of Harm," Arm Resilience Directorate, accessed 17 March 2021.
[86] "Army Sexual Harassment/assault Response and Prevention Continuum of Harm."
[87] "Army Sexual Harassment/assault Response and Prevention Continuum of Harm."

ture and climate of Fort Hood that illustrate the relationship between hegemonic masculinity and sexual assault and harassment.

Observation 1

There is an apprehension to address the relationship between hegemonic masculinities, gender integration, and sexual harassment and assault within the Army, even by an independent review committee established to find answers. FHIRC members immediately took an ambiguous stance addressing the relationship between gender integration and sexual harassment and assault within the context and purpose section with this statement:

> To be clear, this Report does not suggest—and, the Committee has not identified—a direct correlation between sexual harassment and sexual assault and the Army's endeavors toward gender inclusion. However, in reviewing the atmosphere at Fort Hood as it relates to sexual harassment and sexual assault, the Committee is not oblivious to the context of gender integration in the Army.[88]

While the FHIRC does not suggest a direct correlation between sexual harassment and assault and gender inclusion efforts, the committee asserts that a culture and climate that fosters a commitment to inclusion and diversity, freedom from sexual assault and harassment, and adherence to the Army Values is critical to achieving successful gender integration.[89] However, the contents and findings of the FHIRC report reflect a culture and climate in complete contrast to what the committee deemed necessary for successful gender integration. The FHIRC failed to, at a minimum, recommend that culture and climate, sexual harassment and assault, and gender integration within the Army be examined in greater detail. Several observed behaviors and norms within the report's findings indicate that hegemonic masculinities are adversely affecting gender integration and efforts to combat sexual harassment and assault. These problematic behaviors and norms are described and analyzed throughout this chapter.

As far back as 2014, Fort Hood was identified as a high-risk installation for sexual assault and gender hostilities against women by Rand's Workplace and Gender Relations Surveys.[90] Subsequent Rand surveys in 2016 and 2018 confirmed that a dangerous environment for women existed at Fort Hood. In all three surveys, Fort Hood was classified as having the highest risk of sexual assault against women and female gender discrimination. High levels of supervisor workplace gender discrimination against women paired with low levels of peer respect and cohesion were reported on all three Rand surveys over the four-year period. Additionally, the Rand

---

[88] *Report of the Fort Hood Independent Review Committee* (Washington, DC: U.S. Department of the Army, 2020), 1.
[89] *Report of the Fort Hood Independent Review Committee*, 2.
[90] *Report of the Fort Hood Independent Review Committee*, 23–24.

survey reflected low levels of bystander support to intervene and respond responsibly to incidents of sexual harassment and assault, gender hostilities, and discrimination against women.[91]

Rand's studies aim to understand the detail and frequency of sexual assault and harassment within the military. The studies do not attempt to understand the role of culture in creating opportunities for sexual harassment and assault to occur. Further, the studies did not seek to understand why men, in particular, are the primary perpetrators of sexual assault and harassment. This fact should have piqued the curiosity of Army leaders to explore these issues in greater detail following the completion of the surveys in 2014 and 2016 and begs for a recommendation from the FHIRC to suggest reexamining these surveys to create a greater understanding of the culture driving gender discrimination at Fort Hood.

Nonetheless, these risk inventories clearly described patterns of concerning behavior in how women were being treated by male peers and supervisors at Fort Hood as gender integration was in its infancy. The climate concerns also indicated a pattern of dominating and aggressive behavior exhibited by men against women, which is indicative of the pursuit of hegemony or dominance. Because men enacted the majority of these patterns of behaviors, this survey data indicates existing social mechanisms of hegemonic masculinity within the culture and climate of Fort Hood before gender integration. Connell and Messerschmidt note that pursuit of dominance paired with normalizing violent, abusive, and dehumanistic behavior from men and boys within a culture is the primary driver of harmful social mechanisms associated with hegemonic masculinity.[92] Recall that hegemonic masculine ideals exist at the basic assumption level of culture, and basic assumptions are often automatic, meaning we are not aware of the influence they hold over our thinking and behaviors. Often men and boys' aggressive and violent behavior and language is dismissed as "just locker room talk" or by the age-old saying "boys will be boys," which allow unhealthy social mechanisms to continue without intervention. While it cannot be substantiated whether these patterns were conscious or unconscious, research indicates hegemonic masculinity was likely at play.

Further, drawing from Schein's primary embedding mechanisms, leaders integrate the Army Values and tenets of the SHARP Program into organizational culture through what they pay attention to; how they react to crisis and allocate resources; and through deliberate role modeling, teaching, and coaching.[93] These troubling patterns of behavior warranted more direct attention and resources from leaders. Unfortunately, the failure to directly confront these troubling patterns of behavior in the culture of Fort Hood condoned the normalization of deviant behavior within the informal culture.

---

[91] *Report of the Fort Hood Independent Review Committee*, 24.
[92] Connell and Messerschmidt, "Hegemonic Masculinity," 851–52.
[93] Schein, *Organizational Culture and Leadership*, 183.

A similar pattern persisted within the combat brigades at Fort Hood because senior installation leaders ignored the patterns of gender discrimination, sexual harassment, and assault. The FHIRC found the combat brigades within both 3d Cavalry Regiment and 1st Cavalry Division were struggling with promoting a climate of dignity and respect specifically toward women. Within both 3d Cavalry Regiment and 1st Cavalry Division, men outnumbered women by a 7 to 1 ratio, and junior enlisted women comprised most victims of sexual assault and harassment perpetrated by male soldiers.[94] Within Fort Hood, the rates of violent sex crimes were 30.6 percent higher than U.S. Army Forces Command (FORSCOM) averages and 43.2 percent higher than the Army.[95] In 2018 and 2019, sexual assault rates within Fort Hood were noticeably higher than in previous years, and data clearly identified young, junior enlisted female soldiers at high risk of sexual assault. Further, 3d Cavalry Regiment had the highest rate of sex crimes at the time of the investigation, with incidents increasing by 18.6 percent from the previous quarter of the fiscal year (FY) 2020.[96]

Confidential interviews conducted with women in both 3d Cavalry Regiment and 1st Cavalry Division confirmed a culture tolerant of disrespect and abuse toward women. Several women held the belief that the Army only wanted women in combat units for show.[97] This daily negative treatment of women contradicts the Army Values, causing further loss of trust and a sense of exclusion for these women. Many women cited NCOs openly sexually objectifying young female soldiers within their care. Observed behaviors ranged from male NCOs and peers running betting pools to see who could sleep with women first, male NCOs openly stating to an entire unit that "women are here for our entertainment," to male NCOs openly discussing young female soldiers in sexually graphic terms.[98] A senior female NCO told the committee that "sexual harassment and/or assault are 'almost like an initiation to Fort Hood', " and she believed "sexual harassment happens every single day. . . . Nobody stops it; leaders turn a blind eye or they themselves are the offenders."[99] The interviews further confirmed an absence of primary embedding mechanisms; leaders were not paying attention to and addressing behaviors that did not align with a professional work environment within the SHARP Continuum of Harm and were not role-modeling the behaviors they wanted to see.

Commentary by junior enlisted women confirmed this narrative. Particularly within 3d Cavalry Regiment, women reported a disregard for their safety and privacy. Young women living in the post barracks reported NCOs barging into their rooms without notice, often when they were partially dressed.[100] In one instance, an

---

[94] *Report of the Fort Hood Independent Review Committee*, 19.
[95] *Report of the Fort Hood Independent Review Committee*, 21.
[96] *Report of the Fort Hood Independent Review Committee*, 111.
[97] *Report of the Fort Hood Independent Review Committee*, 108.
[98] *Report of the Fort Hood Independent Review Committee*, 108.
[99] *Report of the Fort Hood Independent Review Committee*, 108.
[100] *Report of the Fort Hood Independent Review Committee*, 109.

NCO attempted to sexually assault a female soldier after entering her room without permission. In another, a young female soldier discovered a particular NCO had forcibly entered rooms multiple times and was reported to the chain of command, but the behavior continued. One young soldier reported two counts of sexual assault to her platoon sergeant, who told her, "you can report it, but nothing will happen," and nothing did happen. Women within 3d Cavalry Regiment and 1st Cavalry Division reported a daily struggle to get through their day peacefully without being relentlessly and aggressively pursued in a sexual manner by male soldiers. The FHIRC even noted, "This type of culture towards women in the Enlisted ranks if not addressed proactively creates breeding grounds for sexual assault."[101]

Throughout the interview sessions, the FHIRC uncovered a climate in which women believed they were not wanted and felt unsafe due to privacy violations. Further, the relentless and aggressive pursuit of women by male peers and the sexual objectification that persisted after women reported the behavior is indicative of the removal of autonomy, meaning the advances were unwelcome and imposed against their will. Removal of autonomy is a social mechanism associated with hegemonic masculine ideals often employed to subordinate women and other subordinate masculinities against their will.[102] Revisiting the SHARP Continuum of Harm presented at the beginning of this chapter, the beliefs and climate experienced by these women completely contradict the basic assumptions and espoused beliefs of U.S. Army culture regarding the SHARP Program and the Army Values.

These damning statistics and narratives indicate a culture and climate hostile to women, especially young women who do not hold positional power of rank and authority. Problematic attitudes and violent and abusive behaviors enacted by men against women were quantified in multiple command climate surveys and confirmed through large samples of interview data. This data indicates that hegemonic masculine ideologies and associated social mechanisms were active within the culture of Fort Hood. Throughout the report, the FHIRC hints at connections between gender integration, male attitudes, and perceptions about women and subordinated males and sexual harassment and assault within the U.S. Army. However, the report falls short by failing to recommend further examination of the relationship between the three variables. Considering the sole purpose of the FHIRC was to review the command climate at Fort Hood, the cultural and climate issues discussed in the previous section are begging for actionable recommendations.

Observation 2

Widespread fear of retaliation, exposure, and ostracism for reporting a SHARP violation indicates the enforcement of hegemonic masculinity. The FHIRC identified that women were often silenced when attempting to report sexual assault and harass-

---

[101] *Report of the Fort Hood Independent Review Committee*, 109.
[102] Connell and Messerschmidt, "Hegemonic Masculinity," 835.

ment, if they even chose to report at all. As discussed in the previous section, several women in 3d Cavalry Regiment and 1st Cavalry Division informed the committee that superiors routinely ignored reports ranging from sexual harassment to full-blown sexual assault during confidential interviews. Survey data collected by the FHIRC and the 2018–19 command climate surveys also indicated that fear of retaliation and ostracism was widespread, especially among women and enlisted soldiers. The FHIRC collected 31,612 survey responses, of which 28 percent of women believed that filing a sexual harassment complaint would result in ostracization; 22 percent believed a reporter would be labeled a troublemaker; and 18 percent of women believed a reporter would be discouraged from moving forward with the reporting process.[103]

The same questions were asked of filing a sexual assault complaint, and the percentage of responses were 27 percent, 20 percent, and 17 percent. Once again, the percentages for these survey questions were higher within 3d Cavalry Regiment and 1st Cavalry Division. Further, approximately 1,112 of 5,942 women (19 percent) did not believe that a sexual assault and/or harassment complaint would be kept confidential by the chain of command. Within 3d Cavalry Regiment, 27 percent of female soldiers felt that a sexual assault and/or harassment report would not be kept confidential.[104]

FHIRC individual interview data revealed that of the 507 females interviewed, 32 percent (164 total) would not be comfortable reporting sexual assault or harassment through the SHARP Program at Fort Hood.[105] Approximately 50 percent of the same group of women were not confident in their commanders or that they would take a SHARP report seriously. Regarding retaliation, 36 percent (184 total) of the women interviewed had witnessed or personally experienced acts of retaliation for reporting sexual harassment and/or assault. The overwhelming majority of women interviewed, approximately 70 percent (355 total), believed that Fort Hood's leadership did not execute the SHARP Program effectively.[106]

The survey data collected across Fort Hood confirms a climate of mistrust surrounding the reporting of sexual assault and harassment and demonstrates a misalignment with the espoused beliefs of Army formal culture established at the beginning of this chapter. Per the SHARP Continuum of Harm, leaders are to encourage and facilitate a climate that supports the free and uninhibited reporting of sexual assault and harassment. The FHIRC uncovered several beliefs held by many women at Fort Hood that misaligned with this directive. Women clearly believed that reporting sexual harassment and assault will result in more marginalization, embarrassment, and stress. Most importantly, women did not believe reports of sexual harassment and assault would be taken seriously.

---

[103] *Report of the Fort Hood Independent Review Committee,* 28–30.
[104] *Report of the Fort Hood Independent Review Committee,* 30.
[105] *Report of the Fort Hood Independent Review Committee,* 36.
[106] *Report of the Fort Hood Independent Review Committee,* 36.

The beliefs and narratives surrounding reporting sexual assault and harassment drove soldiers, both men and women, to not report sexual harassment and assault. Through individual and group interviews and the installation-wide survey, the FHIRC confirmed that sexual assault and harassment at Fort Hood were grossly underreported. During interviews with 507 female soldiers, FHIRC discovered 93 counts of sexual assault and 135 instances of sexual harassment.[107] Only 59 of 93 accounts of sexual assault were reported through the SHARP Program. Subsequently, only 72 of the 135 incidents of sexual harassment were reported.[108] The results of the FHIRC installation-wide survey further confirmed widespread underreporting of sexual harassment and assault. Of the 31,000 responses, 1,339 respondents indicated personally witnessing sexual assault within the last 12 months, and 2,625 respondents indicated observing sexual harassment.[109] The results of the FHIRC report starkly contrasted with the cases of sexual assault and harassment known by the SHARP Program at Fort Hood, which recorded 336 counts of sexual assault from 2019 through August 2020. Only 71 reports of sexual harassment were filed from 2019 through August 2020.[110]

The FHIRC uncovered evidence of existing norms, further promoting the silence of victims of sexual assault and harassment. In multiple interview sessions, NCOs revealed the belief that adjudicating sexual harassment and assault was within their realm of responsibility. Further, NCOs exhibited the belief that leadership needed to be shielded from SHARP issues.[111] This dynamic was prevalent within 3d Cavalry Regiment, where approximately 131 male NCOs within the ranks of E5 through E6 expressed a preference and regular practice of informally resolving sexual harassment instead of reporting issues to SHARP personnel for adjudication.[112] While this practice could potentially originate from the common practice of NCOs handling business at the lowest level and could potentially be unconscious, the dynamic contradicts the espoused belief of U.S. Army formal culture and the SHARP Continuum of Harm that charges leaders to report SHARP violations to qualified victim advocates (VAs) and Sexual Assault Response Coordinators (SARCs).

Further, this practice exposes a power dynamic of male leaders preventing women and potentially some men from reporting sexual harassment and assault. Consciously or unconsciously, by handling SHARP complaints at their level rather than allowing VAs and SARCs to handle complaints, these young male NCOs removed the autonomy of the young men and women within their care, taking away the power of victims to seek formal justice and adjudication to complaints. Essentially, young male NCOs took it upon themselves to decide how, if, and on what terms sexual misconduct was handled, forcing women and subordinated masculinities to adapt to

---

[107] *Report of the Fort Hood Independent Review Committee*, 43.
[108] *Report of the Fort Hood Independent Review Committee*, 43.
[109] *Report of the Fort Hood Independent Review Committee*, 44.
[110] *Report of the Fort Hood Independent Review Committee*, 44.
[111] *Report of the Fort Hood Independent Review Committee*, 18.
[112] *Report of the Fort Hood Independent Review Committee*, 41.

the environment imposed on them. This social mechanism is indicative of hegemonic masculinity because by removing victims' autonomy to report sexual harassment and assault, these young NCOs maintained power over those within their care.

All too often, the FHIRC discovered through group interviews that leaders were ignoring and improperly adjudicating instances of sexual harassment and assault. As previously discussed, NCOs within 3d Cavalry Regiment often chose to address sexual assault and harassment reports instead of allowing the reports to be adjudicated by SHARP professionals per Army policy. Often, the FHIRC found leaders were perpetrating sexual harassment and assault, which is itself a behavior that erodes trust in the SHARP reporting system and the leadership at Fort Hood, especially within 3d Cavalry Regiment and 1st Cavalry Division.

Observation 3

The influence of hegemonic masculinity on Army culture is preventing a large portion of male soldiers from understanding how and why culture needs to change. Male NCOs and leaders within the combat brigades at Fort Hood often downplayed the magnitude of the sexual harassment and assault there, which indicates the influence of hegemonic masculinity on the culture of Fort Hood. The FHIRC conducted interviews with 131 junior male NCOs in 3d Cavalry Regiment. During the interviews, most NCOs expressed that they had no concerns about sexual harassment and assault, nor did they take any responsibility or acknowledge the prevalent issues at Fort Hood or within the Army.[113] Interviews with 48 senior male NCOs (E7–E9) within 3d Cavalry Regiment revealed a similar belief that Fort Hood did not have sexual harassment and assault issues, with many expressing that Fort Hood does a better job taking care of soldiers regarding sexual harassment and assault than most colleges.[114]

It is important to note that Specialist Vanessa Guillén was assigned to 3d Cavalry Regiment. In the fallout following Guillen's murder and the accusations of sexual harassment that sparked a national conversation about sexual assault and harassment in the military, and considering the FHIRC report statistics of systemic sexual assault and harassment, these leaders still did not consider sexual assault and harassment a problem. The inability to see sexual harassment and assault as a major problem is likely a side effect of hegemonic masculinity's favorable social conditions, also known as benefits from a patriarchal gender system.[115]

Connell and Messerschmidt identified that dominant forms of masculinity often result in the construct of a patriarchal gender system. All men and even women who enact subordinate masculinities position themselves to receive benefits of this patriarchal system.[116] A common analogy used to define this patriarchal system is the "boys

---

[113] *Report of the Fort Hood Independent Review Committee*, 41.
[114] *Report of the Fort Hood Independent Review Committee*, 42.
[115] Connell and Messerschmidt, "Hegemonic Masculinity," 832.
[116] Connell and Messerschmidt, "Hegemonic Masculinity," 832.

club." Men who do not embody hegemonic masculinity and even women often strive to be a part of the boys club because of the social status and benefits that come with club membership. A potential explanation of why the male soldiers in 3d Cavalry Regiment struggled to see the magnitude of the sexual assault and harassment problem at Fort Hood is that these men unconsciously minimized the problems with sexual assault and harassment at Fort Hood to maintain their achieved social status as a member of the boys club, or to maintain complicity with the system in hopes of gaining social credibility. Speaking out against sexual assault and harassment by publicly refuting a norm of locker room talk or objectification of women in the workplace would likely result in ostracism from the boys club, especially if the norm being refuted is a preferred social mechanism of the dominant group.

An alternate explanation of why young leaders within 3d Cavalry Regiment soldiers struggled to understand the magnitude of the sexual harassment and assault problem at Fort Hood could be because senior leaders were inconsistent in employing primary embedding mechanisms to drive the tenets of the SHARP Program to the lowest levels. When leaders pay attention to too many things or their pattern of attention is inconsistent, subordinates will often use other signals or their own experiences to determine what is important and develop their own set of basic assumptions on how to understand and manage themselves within organizational culture.[117] The commentary from more-senior leaders within the organization suggests that both hegemonic masculinity and leadership failures to consistently apply primary embedding mechanisms contributed to this dynamic.

The most senior male leaders (WO1–O4) interviewed by the FHIRC expressed mixed views acknowledging the problem of sexual harassment and assault at Fort Hood. Some of the leaders expressed that they did not believe Fort Hood was safe for junior enlisted female soldiers, especially in the barracks. However, most of the group did not believe Fort Hood's issues were different from the rest of the Army. The group expressed concerns about professionalism and articulated that they were doing what they could to educate soldiers on appropriate conduct but struggled to understand what constituted "unwanted" behavior in the workplace. Many leaders stated they take the responsibility of mandatory reporting of sexual assault and harassment seriously but felt that many junior enlisted soldiers do not trust field grade leaders because they witness field grade officers committing misconduct.[118]

Following suit to the groups of NCOs, the majority of the officers minimized the problem of sexual assault and harassment with the false justification that Fort Hood was like the rest of the Army. Data showed Fort Hood having the highest rates of sexual assault and harassment within FORSCOM. However, the officers also publicly acknowledged their role in setting the standards for acceptable conduct. Both of

---

[117] Schein, *Organizational Culture and Leadership*, 189.
[118] *Report of the Fort Hood Independent Review Committee*, 42.

these observed behaviors likely coalesce to the preferred social mechanisms of the hegemonic masculine ideals within the officer corps at 3d Cavalry Regiment.

As officers, not publicly acknowledging on some level their roles and responsibilities to sustain a professional work environment and support the SHARP Program would be damaging to their social status as leaders. Further, the officers acknowledged a lack of trust in field grade leadership due to "other" field grade officers committing sexual misconduct around junior enlisted soldiers. The FHIRC did not expand on this comment; however, this comment is indicative that the majority of men believe the underreporting of sexual assault and harassment is primarily because of the individual behavior or a few "bad eggs" rather than understanding that attitudes, beliefs, and unit climate and culture are the primary drivers of sexual assault and harassment.

This demonstrates a cultural misalignment with the SHARP Continuum of Harm discussed at the beginning of this chapter. The officers expressing a lack of understanding regarding what behaviors were unwanted or inappropriate also indicate that a disconnect exists between what the leaders were expected to know about formal culture regarding the SHARP Program and how that knowledge failed to be applied at the informal level of culture.

Across several levels of leadership, male soldiers struggled to acknowledge the magnitude of the sexual assault and harassment problem at Fort Hood and to understand the real challenges faced by female peers. They did not acknowledge their responsibility, nor were they aware of their power to change the situation. This reinforces the previous notion that favorable social conditions created by hegemonic masculinity are unconsciously preventing men from seeing how and why culture needs to change and that men are the primary drivers to make positive changes to make conditions better for everyone. Further, the failure of senior leaders to consistently drive the tenets of the SHARP Program to the lowest levels using primary embedding mechanisms may offer some explanation as to why so many of these leaders developed alternate explanations not necessarily grounded in truth to justify and understand the magnitude of sexual harassment and assault at Fort Hood.

The narrative during mixed-gender group interviews with soldiers from 1st Cavalry Division and 3d Cavalry Regiment further indicates the influence of hegemonic masculinity's favorable social conditions or boys club mentality on the culture of Fort Hood. On multiple occasions, female soldiers would speak up during group interviews to share experiences or flaws with the SHARP Program, only to be undermined, contradicted, and sometimes ridiculed by male members of the group.[119] Male soldiers were comfortable publicly demonstrating hardened attitudes toward female peers in the presence of outside investigators and a Judge Advocate General officer, both of whom were recording transcripts of the session. The vast majority of male

---

[119] *Report of the Fort Hood Independent Review Committee*, 41.

soldiers believed that Army culture did not need to change and that women need to adjust to a male-dominated culture since they chose to join the Army.[120]

The social mechanisms of male soldiers actively undermining female peers could be enacted by the men to prove their masculinity and social status to the other men in the room and FHIRC while simultaneously reinforcing that women speaking out did not have the social status to do so. This dynamic nests with Connell's assertion that the struggle for hegemony or dominance is the true harm of hegemonic masculinities as it can lead to the aggressive behaviors the FHIRC witnessed in the group interviews.[121]

Further, many men publicly expressing the belief that women need to assimilate to the male-dominated culture indicates that the men believe women need to play along with the cultural rituals and norms of the boys club or social conditions of hegemonic masculinity. Consciously or unconsciously, this type of behavior sends the message that if women want to be a part of the team, they must tolerate problematic norms and behaviors within the culture and even sexual harassment and assault. This dynamic enforces compliance among women while allowing hegemonic masculinity's norms and cultural rituals to escape scrutiny.[122] In other words, the problematic attitudes and belief systems that create the climate for sexual harassment and assault in Army culture will only be required to change when men decide to change it. This speaks to the power of hegemonic masculinity within Army culture. What prevents Army leaders from seeing the signs and symptoms of sexual harassment and assault, such as hostile work environments to women, minorities, and some men?

Observation 4

Soldiers who do not feel physically and psychologically safe will not report abusive behavior to superiors, causing leaders to underestimate the scope of the problem with sexual assault and harassment at Fort Hood. While answering the primary research question of this thesis, it became clear that soldiers at Fort Hood, especially female soldiers, did not feel physically safe within their work and living spaces. Through multiple platforms, women informed the FHIRC that reports of hostile work environments and sexual harassment and assault were often blatantly ignored, and if actioned, the women who filed reports faced retaliation and ostracism by peers and superiors. Often, soldiers told the FHIRC they felt physically safer in Kuwait and Afghanistan than at home or work during their service at Fort Hood.[123] Of important note, if Congress had not mandated the review of Fort Hood, none of this information would have ever come to light.

The FHIRC identified that leaders at Fort Hood allowed mission readiness to overshadow integrating the elements of the SHARP Program to the lowest levels.

---

[120] *Report of the Fort Hood Independent Review Committee,* 41.
[121] Connell and Messerschmidt, "Hegemonic Masculinity," 834.
[122] Connell and Messerschmidt, "Hegemonic Masculinity," 835.
[123] *Report of the Fort Hood Independent Review Committee,* 106.

Leaders did not view the SHARP Program as a critical tool to promote soldier safety and morale and to foster a climate of dignity and respect. Additionally, the number of leaders who chose to ignore their responsibility to report sexual assault and harassment through the SHARP Program or who committed acts of sexual misconduct themselves further degraded trust and confidence in both the SHARP Program and within the ranks.[124]

Many soldiers, especially women, at Fort Hood were in survival mode, constantly fearing for physical safety, and they expressed hopelessness in having a safe place to report abusive behavior. Fort Hood's leaders failed to provide a safe working environment that resulted in extreme underreporting of sexual assault and harassment. This made an already dire situation at Fort Hood even worse because a large number of abuses were hidden from view.

While the SHARP Program certainly has flaws, the Department of the Army Inspector General (DAIG) Special Interest Inspection of the Army SHARP Program conducted in 2014 found that commanders who strive to implement the core elements of the program to the lowest levels and take personal ownership of promoting climates of dignity and respect in their units on a daily basis have consistently demonstrated success in reducing—even eliminating—sexual harassment and assault.[125] This research is consistent with the basic assumptions and espoused beliefs of the Army formal culture regarding the SHARP Program, as illustrated using the SHARP Continuum of Harm. Leaders that employ primary embedding mechanisms to integrate the Army Values and tenets of the SHARP Program into organizational culture through what they pay attention to; how they react to crisis and allocate resources; and through deliberate role-modeling, teaching, and coaching sustain a professional work environment and reap the benefits of happier and healthier units.[126] How can Army leaders eliminate behaviors that contribute to sexual harassment and assault and erode trust within the organization?

Observation 5

The research conducted by the FHIRC indicates leaders view sexual assault and harassment as an unsolvable problem, creating a numbness to the occurrence of sexual assault and harassment within the force. To improve U.S. Army culture, leaders must realize their power and authority to positively impact and shape culture.

In answering the primary research question of this thesis, it became clear that social norms existed at Fort Hood that created a permissive climate of sexual harassment and assault and rampant disrespect to women. In the previous sections, female soldiers communicated to the FHIRC that they were openly and aggressively approached in a sexual manner and objectified by fellow male soldiers daily. Further, the women

---

[124] *Report of the Fort Hood Independent Review Committee*, 18.
[125] *Report of the Fort Hood Independent Review Committee*, 31.
[126] Schein, *Organizational Culture and Leadership*, 183.

articulated reporting this behavior to leadership only for leaders to tell them they have no way to stop it. The FHIRC shared the interview dialogue with the senior installation commander at Fort Hood, who quickly responded, "What can I do about it?"[127]

If the senior installation commander at Fort Hood believes that he does not have the agency to address these problematic behaviors, one could speculate this sentiment likely adversely affected subordinate leaders' confidence in their authority to correct the behavior. Certainly, leaders should not set an expectation for women that they should expect to be sexually harassed by male peers at work. Dismissing one's power as a leader to address and correct behaviors that violate the SHARP Continuum of Harm, such as objectification, catcalling, and excessive flirting, is a form of victim-blaming. The statement made by the senior installation commander unconsciously blames the presence of women in the military for systemic sexual harassment and assault rather than the lack of discipline and order that far too often creates the breeding grounds for sexual harassment and assault. Army leaders most certainly can influence the discipline and order necessary to foster a healthy and safe workplace free of sexual harassment and assault. As discussed in the previous section, the DAIG found that commanders who used primary embedding mechanisms to drive the tenets of the SHARP Program and the Army Values into the climate and culture of their units demonstrated success in preventing and reducing sexual harassment and assault.

Further, research by Treviño and Nelson states that socialization, the process of learning how an organization does things, can make or break the sustainment of ethical organizational culture. The broad theory of socialization identifies that generally, people behave in ways consistent with cultural norms because they are expected to, especially in social settings such as work. Further, internalization is when individuals adopt cultural standards as their own.[128] Socialization and internalization are important in understanding ethical and unethical behavior because employees can be socialized into behaving unethically, especially when they do not have the life experience to know the difference between ethical and unethical behavior.[129] When leaders do not correct aggressive, abusive, and violent behavior of men, such as hypersexuality, dehumanistic language, and objectification of women, those behaviors become ingrained as accepted norms and standards of conduct within an organization's informal culture. In the case of Fort Hood, the normalization of hypersexual behavior and language within the informal culture prevented even the most senior leaders from understanding the risk for sexual harassment and assault to the Army's culture and led to their failure to address that risk directly.

Further, leaders' ignorance of patterns of hypersexual behavior, objectification, and use of dehumanistic and aggressive language toward women within the informal

---

[127] Schein, *Organizational Culture and Leadership*, 109.
[128] Treviño and Nelson, *Managing Business Ethics*, 158.
[129] Treviño and Nelson, *Managing Business Ethics*, 158.

culture of Fort Hood indicates the influence of hegemonic masculinity. Often, the dismissal or ignorance of aggressive and violent behavior by men within organizations allows systemic issues of sexual harassment and assault to pervade.[130] For example, young men sexually objectifying women is often dismissed as locker room talk or through the boys will boys mentality. Leaders must recognize the danger in allowing the visible mechanisms of hegemonic masculinity to go unchecked. Both men and women certainly deserve to be held to higher standards of behavior in a professional working environment.

Observation 6

The SHARP Program's primary goal is to address culture change that facilitates discipline and respect; however, addressing attitudes and beliefs and sexual assault and harassment is largely ignored as a prevention strategy. A notable trend the FHIRC report identified is that many leaders still view sexual harassment and assault as individual behavior and climates that are permissive of sexual harassment as isolated incidents. The FHIRC substantiated this assertion noting that the primary focus of the SHARP Program at Fort Hood was response to incidents and victim support rather than emphasizing prevention, which undermined the efficiency of the program at large.[131] This finding explains why so many leaders failed to understand the SHARP Continuum of Harm and the role of attitudes and beliefs underpinning culture and climates that can lead to sexual misconduct. Prevention is undoubtedly separate from response to sexual harassment and assault and requires dedicated time and attention. The FHIRC noted that effective prevention requires data-informed modification of cultural norms to improve group dynamics and social mechanisms by first acknowledging attitudes and beliefs that promote instances of sexual harassment and assault.[132]

In previous sections, it was noted that the FHIRC documented multiple incidents within mixed-gender group interviews where men openly used language to ridicule, contradict, and undermine the concerns female peers had regarding the SHARP Program at Fort Hood. Suppose these men held the basic assumption that women do not matter in the same way that men do, which drove them to disrespect female colleagues publicly. The basic assumption in this instance is problematic, but can Army leaders force someone who holds this belief to change it? Perhaps, but it would take a significant amount of time and buy-in from the individual who holds this basic assumption to change it. A more powerful way Army leaders can foster a climate of prevention is to strictly enforce zero tolerance of disrespectful and dehumanstic language of any kind within the workplace. This strategy sends a clear message that espousing disrespectful attitudes and beliefs is unacceptable. While Army leaders cannot control and change attitudes and beliefs that soldiers carry with them from

---

[130] Connell and Messerschmidt, "Hegemonic Masculinity," 834.
[131] *Report of the Fort Hood Independent Review Committee*, 52.
[132] *Report of the Fort Hood Independent Review Committee*, 52.

childhood, they do have the power and authority to police aggressive, dehumanistic, and disrespectful language. Army leaders can implement this tactic immediately to drive ethical social norms, cultures, and climates. This direct approach to prevention will likely yield far greater success than the compliance-based approach focused solely on executing check-the-block SHARP PowerPoint trainings that fail to drive tenets of the SHARP Program into the culture of units.

Unconscious Practices Maintain Hegemonic Masculinity

In answering the primary and secondary research questions of this thesis, clearly, a pattern of disrespect and aggressive and violent behaviors enacted by male soldiers toward women exists at Fort Hood. This pattern of disrespect and aggression toward women was quantified within surveys that predated the FHIRC investigation and was confirmed within the large samples of survey and interview data collected by the FHIRC. This artifact alone is indicative of the influence of hegemonic masculinity within the culture and climate of Fort Hood and requires further examination across broader Army culture.

The influence of hegemonic masculinity was indicative through beliefs and norms discovered by the FHIRC through group and individual interviews. Many women and men across Fort Hood believe that reporting sexual harassment and assault would result in marginalization and stress and that reports would not be taken seriously. This belief drove significant underreporting of sexual assault. The underreporting and silence surrounding sexual assault and harassment was perpetuated by leaders who either ignored or mishandled SHARP complaints or were offenders of sexual harassment or assault themselves, and it further degraded trust in the SHARP Program. Further, male NCOs within 3d Cavalry Regiment report a common practice of adjudicating SHARP reports rather than allowing reports to be handled by qualified victim advocates. This practice, paired with a low-trust climate, actively prevented the reporting of sexual harassment and assault. Further, these practices were likely unconsciously enforcing hegemonic masculinity by men deciding how and on what terms sexual assault and harassment would be handled, forcing women to subordinate to the patriarchal environments imposed on them.

Most notably, across all interviews and all levels of leadership, male soldiers struggled to understand the magnitude of the sexual harassment and assault problem at Fort Hood even when presented with mounting evidence of its pervasiveness. This dynamic indicates the influence of the favorable social conditions of hegemonic masculinity. A potential explanation is that minimizing or downplaying the magnitude of sexual harassment and assault allows men to maintain complicity with the preferred social mechanisms and norms of the patriarchal gender system or boys club. Speaking out against the problematic norms within the Fort Hood culture contributing to sexual assault and misconduct such as objectification of women or locker room talk could result in ostracization from the boys club. The logic follows Connell's construct of hegemonic masculinity, as all position themselves consciously and unconsciously to continue to attain the benefits of hegemonic masculinity.[133] Even the fear of being ostracized may unconsciously sway some men and women from calling out the boys club behaviors and norms. Lastly, none of the male soldiers interviewed took responsibility for the systemic problems of sexual harassment and assault at Fort Hood. This justification among men indicates the influence of hegemonic masculinity because favorable social conditions and benefits created by hegemonic masculinity unconsciously prevent men from seeing how and why culture needs to change. Further, the justification prevents men from understanding their power to facilitate positive cultural changes to make conditions better for everyone.

Hegemonic masculinity and other problematic attitudes and beliefs possibly underpinned observed patterns of sexual harassment and assault at Fort Hood. The FHIRC missed an opportunity when executing this review by not conducting personal interviews with male soldiers to obtain a deeper understanding of attitudes, belief systems, and masculinities within Fort Hood's culture. Such an analysis could have created an understanding of foundational ideologies and behaviors that cause disrespect between genders. Such a deep level of understanding of the relationship between attitudes and beliefs, culture, and climate will be necessary to design future approaches to true violence prevention and character development in the Army.

---

[133] Connell and Messerschmidt, "Hegemonic Masculinity," 832.

Chapter 5
Conclusions and Recommendations

The major finding of this thesis indicates that there is a potential link between the characteristics of hegemonic masculinity and behaviors in the Army that contribute to a culture that allows sexual harassment and assault. More research is required to prove a definitive link. The FHIRC report did not go far enough by failing to recommend further examination of the culture at Fort Hood and the Army as a whole. The FHIRC likely did not have the time or expertise necessary to examine the problems at Fort Hood from the perspective of organizational culture. However, analyzing the FHIRC report through the lenses of organizational culture and hegemonic masculinity brought to light the influence of hegemonic, competitive masculinity on Army culture and warrants greater examination. The patterns of gender discrimination, sexual harassment, and assault documented in the FHIRC research indicate that more research and analysis of culture is required to understand the attitudes and belief systems driving the behavior.

While a direct correlation between gender integration, hegemonic masculinity, and sexual harassment and assault could not be substantiated within the FHIRC report, unit cultures that promote inclusion, freedom from sexual assault and harassment, and adherence to the Army Values are necessary for both men and women to thrive as soldiers and leaders in the Army. The culture at Fort Hood was not conducive to support gender integration, trust, and inclusion. This led to a climate of mistrust in which victims did not trust the leaders to act on the statements or policies put in place to investigate and prosecute these crimes. The culture of Fort Hood reflects the depth of the problems of sexual harassment and assault that have existed within Army culture for decades.

To stop sexual violence, the research findings of this thesis strongly indicate that the Army requires an understanding of hegemonic masculinity's influence on the basic assumptions, attitudes, and belief systems at the deepest levels of organizational culture within the Army. This understanding will bring the productive and counterproductive beliefs, norms, and behaviors associated with hegemonic masculine ideologies to the attention of Army leaders. By creating awareness of the counterproductive and abusive social norms and behaviors driven by hegemonic masculinity, the Army can identify what systemic biases exist within the organization's culture.

A holistic understanding of systemic bias driven by hegemonic masculinity or other problematic ideologies is critical to facilitate the culture change the Army must make to confront organizational patterns of sexual assault and harassment. Without this understanding, all efforts to drive culture change will fail. One could argue that the most significant reason the Army fails at preventing sexual harassment and assault is because of a failure to understand the problematic ideologies embedded in current

Army culture before implementing programs like the former SAPR Program, Equal Opportunity (EO), and even the SHARP Program.

While these programs are all well-intentioned and designed to support victims and respond to sexual assault and harassment incidents, effective prevention requires a separate and dedicated approach. Both the FHIRC report and research findings of this thesis noted that effective prevention requires data-informed modification of cultural norms to improve group dynamics and social mechanisms by first acknowledging attitudes and beliefs that promote instances of sexual harassment and assault.

In summary, the findings of this thesis demonstrate that more research is required to understand the role of hegemonic masculinity and other ideologies that may contribute to patterns of sexual assault and harassment within the Army. Further, the analysis suggests that a deeper understanding of problematic attitudes and belief systems within Army organizational culture will provide the data necessary to understand underlying assumptions within organizational culture that ultimately drive abusive, violent, and dehumanistic behaviors like sexual assault and harassment. Once Army leaders acknowledge and understand these underlying assumptions, that understanding must inform the design of prevention programs focused on reducing and eliminating sexual assault and harassment.

The first major implication of this research is that the Army will continue to struggle with preventing sexual harassment and assault until senior leaders, the majority of whom are male, understand how and why certain aspects of Army culture must change. While women are not exempt from leading or supporting cultural change in the Army, men still outnumber women by a large percentage, and they hold the largest proportion of leadership positions at the brigade level and below. The narrative surrounding sexual harassment and assault in the military and society generally frame the problem as a women's issue rather than a problem for everyone. This narrative is problematic because it is not correct. Men are also victims of sexual assault and harassment. The narrative that it is a women's issue sends a subconscious message to men that they do not need to pay attention to sexual harassment and assault. Men absolutely must pay attention to issues like sexual harassment and assault in order to effectively change the culture at the root level—the level of underlying assumptions about who belongs in the military and what type of person is tough enough to be on the front line.

A stronger way the Army can leverage men's support to combat sexual harassment and assault is to leverage the Army Values and to have leaders reward actions that build trust and promote a climate of physical and psychological safety. By rewarding soldiers for listening to each other, cooperating with and caring for each other, and raising concerns over safety or cultural norms that make them uncomfortable, leaders can create a climate of psychological and physical safety, resulting in a reduction of sexual harassment and assault incidents. Further, a climate of psychological safety will create opportunities for both women and men to more openly discuss counterproductive workplace issues that affect everyone, as well as issues that are unique to women and minorities. This dynamic creates opportunities for both men and women to exercise empathy toward one another and will likely decrease aggressive, abusive, and dehumansitic behaviors such as sexual harassment and assault, because soldiers will be conscious that certain behaviors do not promote a climate for teamwork and mission success.

The second major implication of this thesis is that leaders within the Army have the greatest power to reduce and prevent sexual harassment and assault. Building on the previous implication and supported with the researching findings of this thesis, the most powerful way Army leaders can foster a culture free of sexual harassment and assault is by publicly upholding the norms and behaviors they want to see. Leaders must support soldiers who report or correct offensive behavior or SHARP violations and recognize soldiers who promote and exemplify the Army Values, to include the tenets of the SHARP Program. Leaders who pay attention to subordinates who do the right things to promote a climate of psychological safety will encourage others to do so as well and will create a healthy, professional work environment.

Further, leaders must openly and consistently dispel soldier misconceptions that everyone endorses norms such as offensive, disrespectful, and dehumanistic language. When norms are publicly upheld that contradict Army values and leaders fail to address them, subordinates will not address the behavior either, likely out of fear of being ostracized from the team. This fear causes soldiers to comply with norms that counter the stated Army Values because those soldiers believe the majority of the group is comfortable with the behavior, consistent with the basic theory of socialization presented in chapter 2. To stop this vicious cycle, officers and NCOs must not engage in and must publicly reprimand offensive behaviors and comments to consistently ensure unit culture and climate is consistent with the Army Values. Additionally, leaders must ensure that soldiers who speak up to report SHARP violations or problematic organizational behavior are not ostracized or retaliated against formally through subpar evaluations or being barred from opportunities or informally through social exclusion and isolation or even bullying.

Building on the major finding of this thesis, Army leadership must initiate cultural studies at Fort Hood and other larger installations, such as Joint Base Lewis-McChord, Fort Bragg, and Fort Campbell, focused on understanding the influence of hegemonic masculinity and other ideologies that may be contributing to patterns of sexual assault and harassment. A large-scale cultural study will provide the Army with the data necessary to understand the attitudes and belief systems driving disrespect between genders and expand on the work of the FHIRC. Further, such a study would provide the Army greater insight in understanding systemic cultural issues contributing to sexual assault and harassment. Additionally, it may even serve as an opportunity to identify installations with successful practices to reduce and prevent sexual harassment and assault. Lastly, in designing such a study, a helpful start may be to seek out research studies on sexual harassment and other dysfunctional workplace behavior conducted in comparable, male-dominated industries such as oil and gas, policing, or finance. It could serve as a frame of reference to inform analysis of Army culture and even provide some validated strategies to address systemic issues.

Another recommendation for future study is for the Army to consider employing Rand to develop and conduct unconscious bias assessments throughout the officer, NCO, and junior enlisted career cycle to provide soldiers and leaders an opportunity to see their own specific biases. The majority of Army leaders across all levels are unaware of their implicit bias tendencies when dealing with different genders, races, or sexualities. Integrating unconscious bias testing across the force would provide the Army with two important opportunities. First, it is highly probable that by simply making leaders aware of their biases and the harm of those biases, leaders will adjust their leadership style to overcome their shortfalls. Second, employing Rand to develop and conduct unconscious bias assessments will allow Rand to aggregate the data to understand systemic bias across the entire Army. Once systemic bias is understood, Army leadership and experts can employ a data-informed approach to design holistic unconscious bias training for integration across the Army. This strategy would provide Army with a twofold approach to address unconscious bias at both individual and organizational levels, which gives the organization a greater chance of successful culture change.

Following the release of the FHIRC report on 8 December 2020, the secretary of the Army addressed the media to discuss the major findings of the report. One quote in particular from his speech captured the heart of the challenge Army leaders must face if they wish to grow following the fallout of the tragic events of Fort Hood:

> The tragic death of Vanessa Guillén and a rash of other challenges at Fort Hood forced us to take a critical look at our systems, our policies, and ourselves. But without leadership, systems don't matter. This is not about metrics but about possessing the ability to have the human decency to show compassion for our teammates and to look out for the best interests of our soldiers.[134]

The longstanding mission of the U.S. Army is to deploy, fight, and win the nation's wars by providing ready, prompt, and sustained land dominance by Army forces across the full spectrum of conflict as part of the Joint Force.[135] The Army's mission cannot take priority as long as damaging norms and behaviors associated with sexual harassment and assault and other dehumanistic behaviors infect unit cultures. Leaders must be meaningfully committed to confronting challenges within informal unit cultures. They must persist in long-term efforts to build and sustain a culture of inclusion, diversity, dignity, and respect until this culture is universal across the force. This type of

---

[134] McCarthy briefing.
[135] "The US Army's Vision and Strategy," Army.mil, accessed 26 April 2021.

commitment is required for the Army to not only prevail and win against internal enemies and corrosive behaviors such as sexual harassment and assault but to build and sustain the diversity of thinking and expertise necessary to prevail and win against all enemies of the United States in close combat.

PART 7

Advising with Gendered Perspectives

Bridging the Gap toward a Gendered Perspective in Security Force Advising

by Lieutenant Colonel Natalie Trogus, USMC[*]

Chapter One
Introduction

> I am Major Abdul Rahman Rahmani, an Afghan Army Aviation pilot . . . I served alongside your husband, MAJ Brent Taylor . . . who was shot yesterday by an evil man . . . I remember him saying, "Family is not something. It is everything." You may or may not be aware of some of our cultural differences, but in Afghanistan, family is not everything, for many of us, family are treated as property. Here, a woman cannot express herself fully, either inside or outside the house. Here, most families treat children unfairly. Let me admit that, before I met Brent, even I did not think that a woman and men should be treated equally. Your husband taught me to love my wife Hamida as an equal and treat my children as treasured gifts, to be a better father, to be a better Husban[d], and to be a better man.[1]

For 20 years, Coalition forces have worked side by side with Afghans like Major Rahmani to stabilize and reconstruct Afghanistan.[2] Achieving fair treatment of men and women in society, at home, and work underpins the practical implementation of UNSCR 1325, Women, Peace and Security (WPS) and *Afghanistan's National Action Plan on UNSCR 1325, Women, Peace and Security*.[3] Gender is a social construct and refers to the roles, rights, and responsibilities attributed to being male or female. Gender also considers ethnicity, nationality, race, religion, and sexual orientation. Social attributes are learned through socialization and differ by culture, thus influencing what behavior is expected and permissible for women and men.[4] Gendered experiences of conflict manifest differently for men and women, shaping their needs and abilities

---

[*] The views expressed in this chapter are solely those of the author. They do not necessarily reflect the opinion of Marine Corps University, the U.S. Marine Corps, the U.S. Navy, the U.S. Army, U.S. Army War College, the U.S. Air Force, or the U.S. government.
[1] Abdul Rahman Rahmani, "Dear Mrs Taylor," Twitter post, 5 November 2018.
[2] This paper was written and submitted to the Joint WPS Academic Forum's writing program prior to the U.S. withdrawal from Afghanistan in 2020. It has not been updated to reflect the current situation.
[3] *Afghanistan's National Action Plan on UNSCR 1325, Women, Peace and Security* (Kabul, Afghanistan: Ministry of Foreign Affairs, 2015). UNSCR 1325 is synonymous with the phrase Women, Peace, and Security (WPS).
[4] *Concepts and Definitions: Women, Peace and Security in NATO* (Brussels, Belgium: North Atlantic Treaty Organization, 2019), 9.

to resolve armed conflicts.[5] Gender norms play a significant role in world affairs, from conflict resolution and peace negotiations in Afghanistan, to women and girls being targets of violent extremists in the Islamic State's terror campaign against Yazidis in Iraq, to women engaging as combatants in Indonesia during the Islamic State in Iraq and Syria's (ISIS) terrorism campaign.[6] From both a national security perspective and as a force multiplier, understanding the unique relationships between gender, conflict, and state security are crucial for planning and executing military advising operations. Gendered behaviors also impact military advising operations. Military commanders seeking enhanced solutions to improve post-conflict security, stability, and peace must account for gender perspectives in all aspects of military operations.

Research has shown when women participate in conflict resolution, conflict negotiations, and the peace process, peace agreements are 35 percent more likely to last at least 15 years.[7] Women's participation in peace negotiations increases gender-sensitive peace agreement provisions, better content of peace agreements, and higher implementation rates. However, gender perspectives have often been overlooked by the Resolute Support Mission (RSM) military leadership and advisors in direct support of the Afghan Peace Process. Despite overwhelming evidence, Afghan women continue to be left out of peace negotiations and excluded from meaningful participation in peace-building operations. As the only female negotiator on a 12-member team at a conference in Moscow, Habiba Sarabi expressed her concerns about so few women being included in important decision-making meetings. Through the U.S.-led RSM, the United States participates directly in the Afghan peace process through military support to the Afghan government. The chairman of the Joint Chiefs of Staff (CJCS), General Mark A. Milley, and RSM commander General Austin S. Miller met the Taliban at the negotiating table to discuss the U.S.-Taliban peace agreement.[8] These meetings are an opportunity for U.S. and RSM leadership to demonstrate support and commitment in implementing gender perspectives supporting Afghan women's role in the peace process. Advising the Afghan Security Institutions (ASI) and Afghan National Defense and Security Forces (ANDSF) requires implementation of gender perspectives through senior leadership advocacy, advisor training, and institutionalization resulting in increased military advising effectiveness in support of reconstruction and peace operations. Senior leaders, such as the chairman of the JCS and the RSM commander, are at the highest levels of command to influence and implement UNSCR 1325. Command authority

---

[5] Robert Egnell and Mayesha Alam, "Introduction: Gender and Women in the Military," in *Women and Gender Perspectives in the Military: An International Comparison*, ed. Egnell and Alam (Washington, DC: Georgetown University Press, 2019), 1.

[6] "Roots of National and International Relations," in *Sex and World Peace*, ed. Valerie M. Hudson et al. (New York: Columbia University Press, 2012), 1.

[7] "Women's Participation in Peace Processes," Council on Foreign Relations, accessed 28 October 2021.

[8] Robert Burns, "After Years Fighting Them, Milley Talks Peace with Taliban," *AP News*, 17 December 2020; and "US Gen. Miller Meets with Taliban in Doha," *Tolo News*, 11 April 2020.

influences subordinates' behavior and education and training of advisors serving in the RSM. Advising the ANDSF requires a well-educated and -trained cadre of advisors to develop and professionalize the ANDSF. Lastly, institutionalization of UNSCR 1325 and gender perspectives must be influenced from RSM senior leadership and operationalized at the advisory level through daily briefs, advisor staff products, and interactions between advisors and advisees.

Research Question and Design

The purpose of this thesis is to inform and influence policy makers and military planners on gendered perspectives of military advising during the Resolute Support Mission.[9] How does an advisor's predeployment training and gendered perspective affect the RSM advising mission in Afghanistan? It is beyond the scope of this thesis to cover the history of the U.S. and the North Atlantic Treaty Organization (NATO) mission and the implementation of UNSCR 1325 WPS. Minimal literature exists on first-hand accounts on the implementation of UNSCR 1325 and will explore advisor gender perspectives and the impacts for the U.S. government and its allies seeking solutions to implement UNSCR 1325 in future operations.

Afghanistan is a useful single case study to analyze this problem. This study uses qualitative data of gender advising experience and empirical examples, coupled with existing research on WPS and gender perspectives in the RSM and discusses inherent challenges in the RSM implementing UNSCR 1325. Subnational examples include advisors from the RSM Headquarters, Ministry of Defense, Ministry of Interior, Gender office, and Train, Advise, and Assist Commands. By borrowing and synthesizing from UNSCR 1325, the *United States Strategy on Women, Peace, and Security*, and DOD's *Women, Peace, and Security Strategic Framework and Implementation Plan* (WPS SFIP), this study recommends opportunities for implementing UNSCR 1325 and gender perspectives in the ASI and ANDSF. By highlighting challenges in implementation of WPS at the advisor level, these first-hand accounts of gender advising perspectives will help policymakers, academics, and professional military educators to refine WPS implementation policy, guidance, and research. This study identifies specific challenges to implementing UNSCR 1325 in Afghanistan advising operations and delivers recommendations to future WPS advising efforts conducted by the U.S. government with an end state to institutionalize UNSCR 1325 in U.S. military advisor predeployment training.

Research Methods

This study's initial intent was to focus on RSM advisors' predeployment advisor training and its influence on the advisory mission's conduct. Interview questions were structured to solicit answers on predeployment training, the advisors' responsibilities,

---

[9] The time period covered here is from July 2019 to July 2020.

and their general understanding of the RSM. The purpose was to understand an advisor's professional and educational background, personal experience on advising, and overall advising actions. Interview questions were prewritten and approved by the institutional review board (IRB) and did not contain gender perspectives because of the author's unfamiliarity with gender perspectives before her own deployment and field research. While serving as a gender advisor, through daily engagements and personal investigation into gender perspectives in military operations, it became evident that gender perspectives were a critical missing element in advisor predeployment training and education. Lack of WPS predeployment training impacts advisors' implementation of gender perspectives during advising operations.

This research uses a mixed-methods approach to examine predeployment advisor training and the gendered impacts training had on the advisor's mission. Primary field research draws on the results of 30 semistructured interviews conducted in Kabul from July 2019 to July 2020. Interviews were focused on advisor predeployment training and overall advising requirements. The purpose of these interviews was to gain insight into RSM's predeployment advisor training, identify gendered perspective patterns across advising experiences and practices, and identify opportunities for enhanced advisor training before assuming duties as an advisor. Advisor gender perspectives incorporated into interviews include advisor rank, nationality, military Service, rank, military occupational specialty, and gender.

Selection of Interview Participants

Primary interview research was conducted in Afghanistan July 2019–July 2020.[10] Advisor contact was made verbally and via email to seek participants for the interviews. Upon confirmation of participants' availability, using a questionnaire, the author scheduled follow-up face-to-face interviews to discuss advisor predeployment training, advising roles, responsibilities, and efforts. The selection process considered the availability of advisors willing to participate, given the challenges of different schedules and operational requirements. All 30 advisors provided informed consent and remain anonymous throughout this study. Figures 1–4 outline essential sex and age-disaggregated data (SADD) of participants such as nationality, military Service, rank, and gender. Disaggregated data provides greater insight and evaluation of the differences between men and women, older and younger, and national military Service, as each perspective will be unique to a particular nation. In chapter 4, advisor interview findings are discussed.

As depicted in figure 7.1, seven different nationalities participated in the interviews, demonstrating the diversity of gender cultural perspectives across the RSM. The United States had 19 military and civilian advisors, which was the most significant contribution of advisors (figure 7.2). As depicted in figure 7.3, military

---

[10] An institutional review board (IRB) was approved for the conduct of in-person interviews on 9 September 2019.

rank and civilian pay grade varied across participants. It is worth noting that under the RSM advising command structure, military servicemembers and civilians were the lead advisors, with the contractors in a supporting role.[11] Of the 30 participants interviewed, 19 were men and 11 were women. Except for two advisors, the remaining 28 participants were ministerial advisors.

Figure 7.1. Nationality

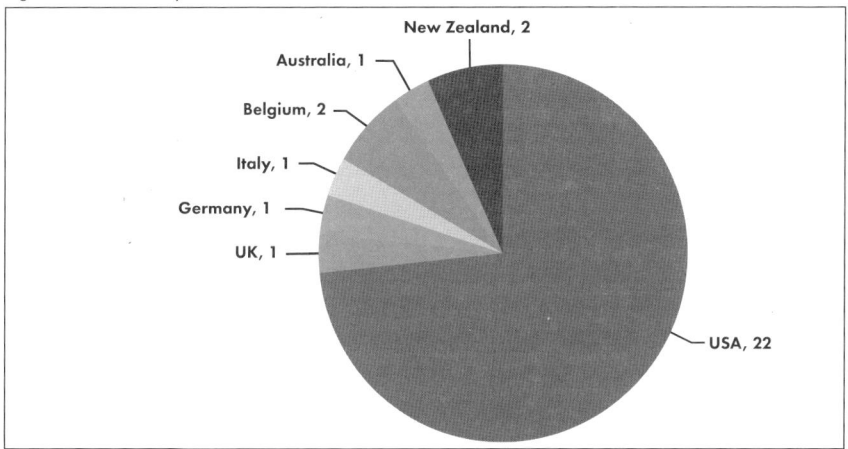

- New Zealand, 2
- Australia, 1
- Belgium, 2
- Italy, 1
- Germany, 1
- UK, 1
- USA, 22

Source: adapted by MCU Press

Figure 7.2. National Military Service

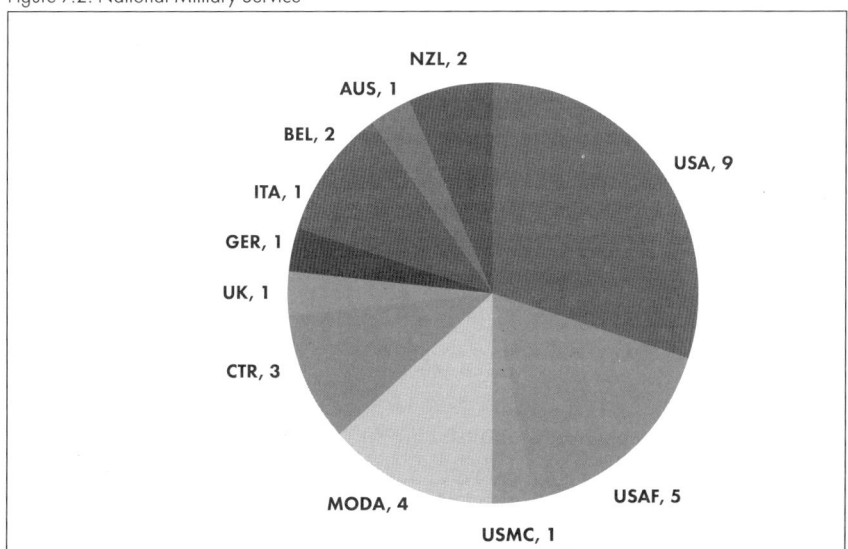

- NZL, 2
- AUS, 1
- BEL, 2
- ITA, 1
- GER, 1
- UK, 1
- CTR, 3
- MODA, 4
- USMC, 1
- USAF, 5
- USA, 9

Source: adapted by MCU Press

---

[11] As expressed as GS: general schedule.

Figure 7.3. Rank

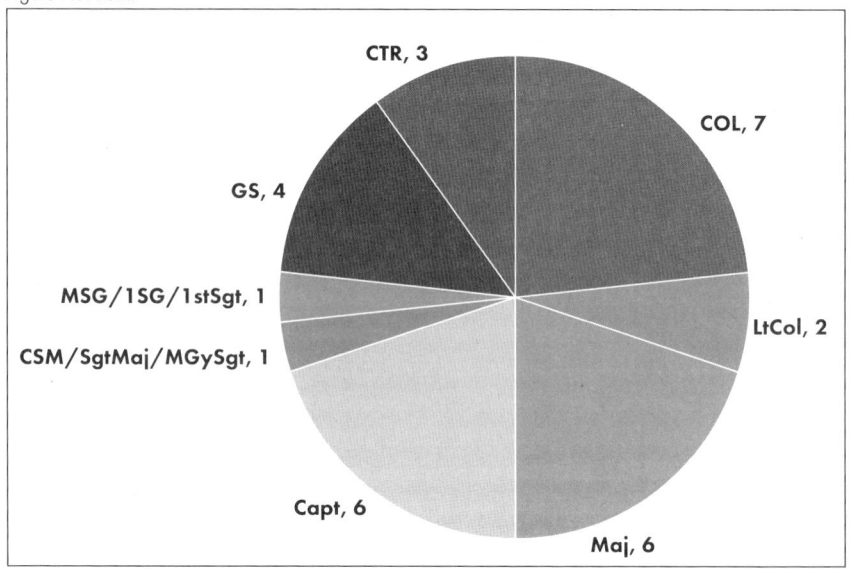

```
CTR, 3
COL, 7
GS, 4
MSG/1SG/1stSgt, 1
CSM/SgtMaj/MGySgt, 1
LtCol, 2
Capt, 6
Maj, 6
```

Source: adapted by MCU Press

Procedures and Limitations

In preparation for the field research, the author drew upon official U.S. government reports including congressional testimonies, Special Inspector General Afghanistan Reconstruction (SIGAR), Enhancing Security and Stability in Afghanistan congressional reports, and Afghan South Asia Strategy. Additional resources included military doctrines such as the *Multi-Service Tactics, Techniques, and Procedures for Advising Foreign Security Forces*; books, journal articles, and newspapers gathered through library search engines; and open-source reporting.[12] As this is a single case study, this research will be focused on the RSM train, advise, and assist advisory mission in Afghanistan from July 2019 to July 2020.

The RSM operational environment during the author's field research was politically sensitive. Faced with a contested Afghan presidential election in 2019, tensions mounted during the fall 2019 fighting season and violence levels escalated leading up to U.S. and Taliban peace negotiations. In September 2019, peace talks abruptly ended and regional tensions increased with Iran and Pakistan. In February 2020, the United States and the Taliban signed a simultaneous peace deal and joint declaration. Domestic tensions surrounding post-peace rising violence levels and rhetoric of U.S. troop withdrawal fueled uncertainty. In March 2020, the COVID-19 pandemic

---

[12] *Afghanistan National Defense and Security Forces: DOD Lacks Performance Data to Assess, Monitor, and Evaluate Advisors Assigned to the Ministries of Defense and Interior* (Washington, DC: Special Instructor General for Afghanistan Reconstruction, 2018); *Enhancing Security and Stability in Afghanistan* (Washington, DC: Department of Defense, 2020); and *Multi-Service Tactics, Techniques, and Procedures for Advising Foreign Security Forces* (Hampton, VA: Air Land Sea Application Center, 2014).

abruptly suspended face-to-face advisor meetings and impacted the ability to conduct face-to-face advisor interviews for this study. Additionally, per the U.S.-Taliban peace agreement, the pandemic accelerated coalition advisor redeployment. A month after the signing of the peace agreement, violence levels continued to rise with targeted assassinations and attacks perpetrated against women and children, including attacks on the Kabul Maternity Clinic and Kabul University.

The following challenges affected field research in Kabul:

1. The release of the Afghanistan Papers, limiting the ability to meet with several advisors.[13] The Afghanistan Papers revealed the names of previously anonymous interviewees. As a result, interview participants became concerned that their personal information and candid comments would be exposed and feared retribution from senior leadership.
2. The COVID-19 pandemic shut down the RSM advisory mission from March 2020 to July 2020, limiting the ability to meet face to face for interviews that were crucial to capturing advisors' experiences.
3. The U.S.-Taliban peace deal and joint declaration resulted in an immediate withdrawal of U.S. troops, resulting in the in March and April 2020 redeployment of advisors who had agreed to participate in this research.

Interview Questions

To obtain insight of advisors' predeployment training and the effects of training on the performance of advising, 35 questions were asked. A few questions are included below. For a complete list of interview questions, see appendix B.

1. What was your source of deployment?
2. What are some of your advising responsibilities?
3. What are the limitations of your advising efforts?
4. Did you receive advisory predeployment training prior to deployment?
5. Did you read any publications on advising prior to deployment?
6. Have you read the RSM operation plan?

Interview questions were structured to remain unbiased and open-ended to solicit candid responses. Responses were analyzed for gendered training, gender awareness, and gender perspectives in the conduct of individual advisor actions. Advisor familiarity with the RSM operational plan (OPLAN) during predeployment training would expose advisors to the NATO military strategic objective of implementing UNSCR 1325 into the RSM and requirement to include gender perspectives into all military operations.

---

[13] "The Afghanistan Papers. A Secret History of the War," *Washington Post*, accessed 11 December 2019.

Findings

To rectify weak leadership, substandard WPS advocacy and absence of WPS institutionalization I identify four recommendations. The first is to conduct a holistic research study into the challenges of implementing UNSCR 1325 at the advisor level. Personal insight will provide valuable feedback to support future policy generation for future advisor employment. Mandatory gender sensitive pre-deployment training is the second recommendation. In order to effectively implement gender perspectives, advisors need to be aware of the subject area and the gender dynamics of Afghan culture to effectively apply Afghan-sensitive gender perspectives. Third, WPS implementation must be driven from the highest levels within DOD. For WPS to take root in U.S. military missions, the Office of the Secretary of Defense designee who is responsible for DOD WPS implementation must be engaged with U.S. mission commanders. Lastly, a gender-sensitive budget execution plan must be developed in order to effectively apply congressional appropriations for gender initiatives. Establishing a gender budget execution unit will facilitate the direct support of gender programs, reduce likelihood of corruption, and prevent money from being ciphoned out of gender programs.

The next five chapters investigate the primary research question of how predeployment training and gendered perspective affect the RSM advising mission in Afghanistan. Chapter 2 provides a background on UNSCR 1325 and discusses gendered aspects of building partner capacity and security cooperation advising efforts, which the coalition has implemented to support Afghanistan's reconstruction efforts. Chapter 3 provides a background on the U.S. response to UNSCR 1325 and the RSM advisory efforts. Through applying a gender analysis and gender perspective, chapter 4 discusses three key findings identified during research and interviews. Despite strategic guidance, inclusion of gendered perspectives at the advisory levels remains limited due to weak leadership advocacy, substandard WPS advisory training, and the absence of WPS institutionalization directly impacting the implementation of UNSCR 1325 and gender perspectives in the RMS. Chapter 5 summarizes the research and offers recommendations to support Afghanistan's future reconstruction efforts and the implementation of UNSCR 1325 and gender perspectives in support of the Afghan national action plan for implementing UNSCR 1325.

Chapter Two
Background on UNSCR 1325, Women, Peace, and Security

The end of the Cold War saw a rise in post-conflict and late-developing states that have required international assistance, often including military intervention. Many post–Cold War states have suffered from internal violence and conflict such as civil wars, foreign invasions, and in recent years terrorist activity. State and government institutions are neglectful of civilian populations, incapable of responding to crisis, and often challenged to meet societal expectations. International interventions often focus on restoring and rebuilding state, defense, and security institutions through capacity-building programs to provide a foundation for the host government to restore services. As assistance efforts reconstruct institutions, women's inclusion brings different perspectives, insights, and mediation styles effective for reconstruction efforts.[14] Research has also demonstrated that women serving on the front lines of effort play a meaningful role in preventing conflict when tension rises in society's social strata.[15] Afghanistan has presented a unique set of multiple and simultaneous post-conflict circumstances the international community must address.

Geopolitically, Afghanistan is a landlocked buffer state between two nuclear-weapon states: Iran and Pakistan.[16] To further complicate matters, Afghanistan shares a border with Pakistan which is in a persistent existential conflict with India, including the threat of nuclear escalation. Afghanistan has experienced many conflicts ranging from regime failure, Soviet occupation, civil war, humanitarian crisis, violent extremist regime, and invasion, which has resulted in its designation as a failed state. After decades of conflict and civil war, in the early 1990s the Taliban arose to stabilize war-torn Afghanistan. As the movement gained momentum in the late 1990s the Taliban's reins of power were built on tyranny, repression, and fear.[17] The Taliban harbored al-Qaeda and as a result of the attacks on 11 September 2001 (9/11), the United States invaded Afghanistan in October 2001 and began an international effort to rebuild the Afghan state.

Historically, Afghanistan has been a rentier state relying on foreign economic assistance that continues to the present time.[18] As one of the pillars of state development

---

[14] Marie O'Reilly and Andrea Ó Súilleabháin, *Women in Conflict Mediation: Why It Matters* (New York: International Peace Institute, 2013), 5.
[15] Marie O'Reilly, "Inclusive Security and Peaceful Societies: Exploring the Evidence," *Prism* 6, no. 1 (March 2016): 21–33.
[16] Barnett R. Rubin, *The Search for Peace in Afghanistan: From Buffer State to Failed State* (New Haven, CT: Yale University Press, 1995), 121.
[17] Hassan Abbas, *The Taliban Revival: Violence and Extremism on the Pakistan-Afghanistan Frontier* (New Haven, CT: Yale University Press, 2014), 61.
[18] Rubin, *The Search for Peace in Afghanistan*, 121; and Astri Suhrke, "Statebuilding in Afghanistan: A Contradictory Engagement," *Central Asian Survey* 32, no. 3 (October 2013): 272, https://doi.org/10.1080/02634937.2013.834715.

and capacity building, economic prosperity is crucial for post-conflict failed states.[19] The international community does not have a blank check by which to continue supporting states undergoing capacity building. There is a fine line on building capacity in these fragile societies while achieving measurable outcomes simultaneously. International partners provide vital economic resources, which, at a cost to the host government, have little incentive to develop their organic capacity.[20] Ethnic divisions also influence effective capacity-building programs.[21] Ethnicity is an underpinning that Western nations often ignore due to historical norms and lack of familiarity with cultural conditions.

Afghanistan fits several doctrinal paradigms in terms of operational approach; the international community commits resources to address state failure, including stability, peacekeeping, reconstruction, counterterrorism, humanitarian, security cooperation, and security assistance operations. In Afghanistan, all these paradigms are simultaneously executed. The theoretical underpinning to help Afghanistan recover after four decades of conflict has been international foreign assistance by conducting capacity building ranging from peace and stability operations to long-term reconstruction and development programs. To address Afghanistan's future needs and raise its status from a failed state to a prosperous state, it is imperative to understand how the United States has prioritized capacity building in failed states, inclusive of gendered perspectives in Afghanistan. Reconstruction efforts inclusive of women's perspectives facilitate the advocacy for policies and procedures that lessen conflict drivers in ways that less gender-diverse groups do not.[22] Groups that exclude women, minorities, and those with disabilities often fail to consider the needs of the population and account for gender-sensitive social programs and resources crucial to increasing social stability and security.

Afghan Women's Rights Under the Taliban

Afghan reforms in the early 1920s and again in the 1950s included ending the seclusion of women and abolishment of the veil. Reforms and social changes were drastic from Kabul and the rest of the country as a high concentration of modernizing institutions and more liberal attitudes of the population was concentrated in Kabul. Reforms were introduced first in Kabul with the intent of extending to the rural areas were gradually applied. Unveiling was seen as a choice versus compulsion and did not take root in the rural areas as women continued to observe strict purdah in con-

---

[19] Peter Tomsen, *The Wars of Afghanistan: Messianic Terrorism, Tribal Conflicts, and the Failures of Great Powers* (New York: Public Affairs, 2011), 709.
[20] Suhrke, "Statebuilding in Afghanistan," 273.
[21] Sven Gunnar Simonsen, "Building 'National' Armies—Building Nations?," *Armed Forces and Society* 33, no. 4 (July 2007): 572.
[22] O'Reilly and Ó Súilleabháin, *Women in Conflict Mediation*, 5–6.

servative areas.[23] Unveiling was seen as a larger project designated to transform Afghan life through the emancipation of women. Reforms directly affecting women included restrictions on the age of marriage, and proof of consent, limiting bride price and wedding prices, and prohibitions on polygamy, which all had direct impacts on Afghan women's rights.[24] Life under the Taliban reverted to strict seclusion, banning from all social spheres, and mandatory veiling.

At the end of the Cold War, the pattern of conflict in Afghanistan restructured society to exclude women from public space. Afghan women have always played a limited role in public politics, however some have served as ministers and members of parliament. Decades of war brought the militarization of society as well as destruction of productive activities and resources women had access to. As resources became scarcer, women were disproportionately affected as males, the only armed gender, gained more access to whatever was available.

As the Taliban movement gained supporters and strength, the primary focus was on the military campaign and the imposition of an obscurantist version of Sharia law that was a mixture of tribalism, male chauvinism, and illiteracy.[25] They forced women into seclusion and imposed brutal forms of punishment to control women such as beatings, stonings, and executions in front of large crowds of men and boys. The Taliban codified existing regulations limiting women's education and economic empowerment and enforced them with harsh punishment. The pervasive male feeling of dishonor as a result of the war led to harsher attempts to control women, who in patriarchal societies are seen as repositories of the honor of their families.[26] When Madeline Albright became secretary of state under the Clinton administration, she was outspoken and was the driving force behind the United States publicly condemning the Taliban's gender policies. Supported by feminist, human rights, and lobbying groups, Taliban gender policies became a political issue.

After capturing Kabul, in 1996 the Taliban issued a decree relating to women and other cultural issues. Specific to women, this decree segregated Afghan men and women in all social settings, restricted women's freedom of movement, and delineated the role of Afghan women in society and in the home. The decree stated that when leaving their residence, women needed to be covered in burqas in accordance with Islamic Sharia regulation.[27] The Taliban encouraged domestic violence toward women

---

[23] Use of the veil fell out of fashion by the women of Kabul during this period. However, women in the countryside continued to wear the veil as it was seen as a social status, following the urban elite who had chosen to wear the veil and its association with not needing to work.

[24] Anand Gopal, *No Good Men Among the Living: America, The Taliban, and the War Through Afghan Eyes* (New York: Picador Books, 2014), 202.

[25] Hassan Abbas, *The Taliban Revival: Violence and Extremism on the Pakistan-Afghanistan Frontier* (New Haven, CT: Yale University Press, 2014), 69.

[26] Barnett R. Rubin, *Afghanistan from the Cold War through the War on Terror* (New York: Oxford University Press, 2013), 31.

[27] Ahmed Rashid, *Taliban: Militant Islam, Oil and Fundamentalism in Central Asia* (New Haven, CT: Yale University Press, 2000), 248.

who violated any aspects of the decree, thus establishing one of the worst forms of gender apartheid in the 21st century.

The Taliban established the Amar Bil Maroof Wa Nahi An al-Munkar, or the Department of the Promotion of Virtue and Prevention of Vice. Also known as the religious police, young men walked around the streets with whips, long sticks and Kalashnikov rifles looking for violations of strict dress codes. In 1997, Afghan women were banned from wearing high heels, making a noise with their shoes while they walked, and from wearing makeup.[28] The contrast was stark, as in the late 1980s Afghan women were attending universities, teaching, and working as doctors, nurses, and professors—some of them in miniskirts, but many of them unveiled.

Gender Inequality and Armed Conflict

Research on gender inequality and armed conflict suggests a connection between women's security and a state's peacefulness.[29] Valerie M. Hudson et al's research linking the security of women to the security of states found empirical evidence linking the two variables.[30] Gender performance varies widely across cultures. Few works recognize the intimate relationship that may exist between the political order between men and women and the political order that develops within the nation-state. Fewer studies contemplate the relationship between these two political orders and the resulting stability and resilience of the nation-state.[31] The idea that the sexual order and political order are linked is not new. Suggesting that political order is based on the structure of male and female relations—that is, the established sexual political order shaping the development of societal collective political order—is of critical importance to military coalitions supporting nation-state stability and reconstruction efforts.

Sexual political order is focused on building fraternity through the systematic subordination of women that uses patrilineality to facilitate the creation of fraternity. Interlocking patterns of institutions, processes, and norms that enforce it is termed Patrilineal/Fraternal Syndrome.[32] For peace, stability, and security to be established, the syndrome must be disrupted and dismantled. Hudson et al. suggests populations in Central Asia and Southeast Asia see the prevalence of clan politics, which propels real-world challenges to governance, economic performance, and in some cases stability.[33] States failing to provide security for their citizens will propel groups to resort to clan-based governance in the absence of authority and is a consideration for future building partner capacity efforts. The most successful form of social control in

---

[28] Rashid, *Taliban*, 105.
[29] Valerie M. Hudson, Bonnie Ballif-Spanvill, Mary Caprioli, and Chad F. Emmett, "The Heart of the Matter: The Security of Women, the Security of States," *Military Review* (May–June 2017): 28.
[30] "The Heart of the Matter: The Security of Women and the Security of States," in *Sex and World Peace*, 110.
[31] Valerie M. Hudson, Donna Lee Bowen, and Perpetua Lynne Nielsen, *The First Political Order: How Sex Shapes Governance and National Security Worldwide* (New York: Columbia University Press, 2020), 11.
[32] Hudson, Bowen, and Nielsen, *First Political Order*, 44.
[33] Hudson, Bowen, and Nielsen, *First Political Order*, 25.

Afghan society is the patriarchal family. It is impossible for Afghan women to escape male domination and is a domestic contention that women fight hard against.[34] It is under this control that Afghan women have self-organized and have created a distinct women's culture, which when given the chance, could enable them to defend themselves and meaningfully participate in conflict resolution, conflict negotiations, and peace building.

Building Partner Capacity

Gender-sensitive building partner capacity efforts in failed states provide a lens to analyze and address root causes of conflict. According to Francis Fukuyama, Barnett R. Rubin, and Thomas Barfield, international assistance in post-conflict states suffering from state failure require international nation-building assistance to develop state capacity.[35] Brinkerhoff defines *capacity* as dealing with aptitudes, resources, and relationships, and facilitating conditions necessary to act effectively to achieve some intended purpose.[36] Capacity can be addressed at three levels: the individual, the organizational, and the institutional. Capacity development is linked between these levels and is mutually dependent on one another. Post–9/11, Afghanistan state building was an extremely challenging endeavor. The U.S. toppling of the Taliban left the country with yet another vacuum of power that needed to be quickly resolved. William Maley found that the rebuilding of Afghanistan as initially conceived at the Bonn Conference in 2001 was not doomed from the outset; rather, it was undermined by a series of strategic misjudgments and miscalculations that combined to produce a dispiriting, rather than inspiring, outcome.[37] State building in failed states requires the rebuilding of new institutions from the ground up. However, the massive requirements and intervention from the international community can have both positive and negative outcomes. Afghanistan was a unique state after the fall of the Taliban. The international community's inherent contradictions led to immense challenges in developing state capacity after the Bonn Conference.[38] Although research on nation building and women does not demonstrate a direct link between military intervention and women's equality, it is essential to note military intervention can support the security situation.[39]

---

[34] Barnett R. Rubin, *The Fragmentation of Afghanistan: State Formation and Collapse in the International System*, 2d ed. (New Haven, CT: Yale University Press, 2002), 41.

[35] Francis Fukuyama, "Guidelines for Future Nation-Builders," in *Nation Building Beyond Afghanistan and Iraq*, ed. Francis Fukuyama (Baltimore, MD: John Hopkins University Press, 2006), 232; Rubin, *Fragmentation of Afghanistan*, 62; and Thomas Barfield, *Afghanistan: A Cultural and Political History* (Princeton, NJ: Princeton University Press, 2010), 293.

[36] Derick W. Brinkerhoff, "Developing Capacity in Fragile States," *Public Administration and Development* 30, no. 1 (February 2010): 67, https://doi.org/10.1002/pad.545.

[37] William Maley, "Statebuilding in Afghanistan: Challenges and Pathologies," *Central Asian Survey* 32, no. 3 (October 2013): 258, https://doi.org/10.1080/02634937.2013.834719.

[38] Suhrke, "Statebuilding in Afghanistan," 272.

[39] Mary Caprioli and Kimberly Lynn Douglass, "Nation Building and Women: The Effect of Intervention on Women's Agency," *Foreign Policy Analysis* 4, no. 1 (January 2008): 45–65, https://doi.org/10.1111/j.1743-8594.2007.00057.x.

Literature on women's participation in peace building and post-conflict reconstruction suggests linkages in improving security allows women to access crucial foreign aid organizations' basic needs, including health services, food aid, and education.

The DOD defines *building partner capacity* as a broad set of missions, programs, activities, and authorities intended to improve other nations' ability to achieve those security-oriented goals they share with the United States. However, the effectiveness of building partner capacity and whether or not strategic objectives have been achieved has been a debated topic among scholars and the U.S. Congress. Since 2001, building partner capacity has been a core component of the U.S. efforts in Iraq and Afghanistan. It only recently has been prioritized in U.S. administrations in addressing weak and failing states.[40] However, it was not until 2006 when building partner capacity was formally discussed in U.S. strategy and foreign policy documents, as addressed in 2006.

> Long-duration, complex operations involving the U.S. military, other government agencies and international partners will be waged simultaneously in multiple countries around the world. ... Maintaining a long-term, low-visibility presence in many areas of the world where U.S. forces do not traditionally operate will be required. Building and leveraging partner capacity will also be an absolutely essential part of this approach, and the employment of surrogates will be a necessary method for achieving many goals.[41]

From a historical U.S. policy standpoint, and before 2001, building partner capacity was focused on security force assistance (SFA) missions, conducting bilateral engagements and training exercises, including military-to-military engagements, and U.S. forces conducting training alongside foreign militaries.[42] The United States began operations in Afghanistan to destroy al-Qaeda for attacking the U.S. homeland and to destroy the Taliban for harboring Osama bin Laden.[43] Combat operations against the Taliban succeeded and resulted in the organization's overthrow on 9 December 2001. Seventeen months later in May 2003, the United States declared an end to major combat operations. However, post-combat was a fragile time for the Afghan state and was a crucial period for state-building initiatives to take place. U.S. strategy and policy lacked focus on Afghanistan reconstruction efforts. Instead, it focused on killing or capturing bin Laden and turned attention toward Iraq in a broader Global War on Terrorism. Still, since 2002, the United States has committed more than $190

---

[40] Kathleen J. McInnis and Nathan J. Lucas, *What Is "Building Partner Capacity?": Issues for Congress* (Washington, DC: Congressional Research Service, 2015), 1, 5.
[41] *Quadrennial Defense Review Report* (Washington, DC: U.S. Department of Defense, 2006), 23.
[42] Kenneth Katzman and Clayton Thomas, *Afghanistan: Post-Taliban Governance, Security, and U.S. Policy* (Washington, DC: Congressional Research Service, 2017), 6–10.
[43] Laurel E. Miller, *The Challenges and the Benefits for U.S. National Security of Providing Foreign Assistance to Afghanistan* (Santa Monica, CA: Rand, 2018), 1–5.

billion in aid and reconstruction in Afghanistan.[44] In January 2015, a new NATO-led mission, Resolute Support, focused on training, advising, and assisting government forces to include defense and security institutions at the ministerial level.[45] In 2017, Trump unveiled his South Asia Strategy which outlined a conditions-based approach to Afghanistan and was reliant on NATO allies and global partners to support the new strategy.[46] The 2017 *National Security Strategy of the United States of America* (NSS) outlines and reinforces the importance of partner alliances and working alongside countries with the same shared values to maintain international order and stability.[47] Partnerships are crucial to U.S. strategy are is one of the lines of effort outlined in the National Defense Strategy.[48] The emphasis on partner alliances reinforces the critical task of building partner capacity and working alongside allies to ensure mutual and collective defense through advising partner nations.

Security Cooperation Advising

Nadia Gerspacher states that advising missions are fundamentally a capacity-building activity that helps restore the host government's ability to govern.[49] The United States has prioritized capacity building through the execution of its security cooperation Title 10 authorities and establishment of the Ministry of Defense Advisor (MODA) program and Afghanistan Pakistan Hands (APH) program, which marked a significant evolution in the DOD approach to capacity building at the institutional level.[50] Very few scholars have considered the implementation of advising strategy and advisor training to execute defense institution building as a capacity development tool and a crucial factor that affects the building of state capacity in failed states. Gerspacher, however, finds evidence that it could be a substantial contributing factor in some cases.[51] Alexandra Kerr and Michael Miklaucic claim defense institution building is an effective method and approach to support failed states in building capacity; however, it is not clear if military advisors from conventional forces are adequately prepared to conduct defense institution building at the strategic level due to competing missions of combat operations and advising and the sustainment and predeployment training involved

---

[44] Clayton Thomas, *Afghanistan: Background and U.S. Policy in Brief* (Washington, DC: Congressional Research Service, 2018), 7.
[45] Katzman and Thomas, *Afghanistan*, 1.
[46] Donald J. Trump, "Remarks by President Trump on the Strategy in Afghanistan and South Asia" (speech, the Pentagon, Arlington, VA, 21 August 2017), 5.
[47] *National Security Strategy of the United States of America* (Washington, DC: Office of the President, White House, 2017), 37.
[48] *Summary of the 2018 National Defense Strategy of the United States of America: Sharpening the American Military's Competitive Edge* (Washington, DC: U.S. Department of Defense, 2018), 8.
[49] Nadia Gerspacher, *Strategic Advising in Foreign Assistance: A Practical Guide* (Boulder, CO: Kumarian Press, 2016), 1.
[50] James Schear et al.,"Ministerial Advisors: Developing Capacity for an Enduring Security Force," *Prism* 2, no. 2 (March 2011): 136; and *Chairman of the Joint Chiefs of Staff Instruction 1630.01B, Afghanistan/Pakistan Hands (APH) Program* (Washington, DC: Joint Chiefs of Staff, 30 May 2019).
[51] Gerspacher, *Strategic Advising in Foreign Assistance*, 1.

in preparing conventional forces for conventional war.[52] It goes without arguing in post-conflict states that a basic set of requirements is needed to assist the state.

UNSCR 1325 Women, Peace and Security

Twenty years ago, on 31 October 2000, the UN Security Council unanimously adopted UNSCR 1325.[53] This landmark security resolution was the first of its kind to formally recognize the disproportionate effect conflict has on women and girls while recognizing the unique perspectives women bring to conflict resolution, conflict negotiations, and peace building. At the time, this resolution was exceptional as it shifted global thinking regarding conflict and its aftermath, bringing to the forefront gender considerations incorporated into a nation's foreign and diplomatic policies. Additionally, this revolutionary resolution increased awareness between gender inequality and international peace and security. It also brought forth discussions on the effects of conflict and treatment in domestic lives of citizens exposed to conflict, such as treating wives and children as equal partners, as highlighted by Major Rahmani's sentiments to Major Taylor's wife.

UNSCR 1325 has three themes, including conflict resolution, conflict negotiation, and peace building. Furthermore, the resolution has four pillars: participation, protection, prevention, and relief and recovery. Ten additional resolutions were adopted, encompassing the WPS agenda.[54] The agenda provides member states a framework for implementing WPS principles into civil, defense, and security sectors and more specifically in the planning and conduct of military operations.[55] The WPS agenda addresses sexual gender-based violence, rape as a weapon of war, and children in armed conflict. These resolutions call for UN member states to protect women and girls from sexual gender-based violence and employ the resolutions in accordance with legally binding laws such as international humanitarian laws and human rights laws. Obligations under the Geneva Conventions and international laws supporting the WPS agenda, which requires parties in a conflict to prevent violations of women's rights, support women's participation in peace negotiations and post-conflict reconstruction.

The WPS agenda calls on member states to integrate gender perspectives into military operations, which includes military advising operations in Afghanistan.[56] NATO has responded to the demands of UNSCR 1325 throughout its organization by integrating gender perspectives across the alliance's core tasks of collective de-

---

[52] Alexandra Kerr and Michael Miklaucic, ed., *Effective, Legitimate, Secure: Insights for Defense Institution Building* (Washington, DC: Center for Complex Operations, 2017), 11.
[53] United Nations Security Council, Resolution 1325 Women, Peace and Security, S/RES/1325 (31 October 2000).
[54] "Women, Peace and Security," North Atlantic Treaty Organization, accessed 28 November 2020.
[55] "The Resolutions," Women's International League for Peace and Freedom, accessed 28 November 2020.
[56] Robert Egnell, "Gender Perspectives and Military Effectiveness: Implementing UNSCR 1325 and the National Action Plan on Women, Peace and Security," *Prism* 6, no. 1 (March 2016): 72–89.

fense, crisis management, and cooperative security. NATO calls on countries to use a gender lens in defense and security sector planning and operations to enhance military effectiveness through engagement of the entire population and to seek solutions to enhance security.

A gender perspective is a way of assessing gender-based differences between women and men as reflected in their social roles and interactions in the distribution of power and access to resources.[57] When advising the Afghan defense sector in the implementation of UNSCR 1325, through *Afghanistan's National Action Plan on UNSCR 1325, Women, Peace and Security*, a gender perspective is essential in establishing gender-sensitive ANDSF professional development programs and ensuring that women's and men's specific needs are included in programming and resource distribution. NATO recognizes the disproportionate impact conflict has on women and girls and the importance gender perspectives have on NATO military operations in critical areas such as intelligence, operations, and decisionmaking.[58] For this reason, the RSM OPLAN included the implementation of UNSCR 1325 as a NATO military strategic objective.

As outlined in the U.S. South Asia Strategy, Afghanistan reconstruction efforts seek to build trust and technical expertise among the ANDSF through building partner capacity programs. The strategy is aimed at creating conditions on the ground to prevent Afghanistan from becoming a terrorist safe haven from which terrorists can launch attacks on the U.S. homeland. After more than 40 years of violent conflict and more than $190 billion dollars, political instability remains. Terrorist activity targets Afghan citizens, resulting in nearly two generations of Afghans growing up in wartime conditions. As a result, violent and aggressive behavior has become status quo in Afghan society, which negatively impacts Afghan gender roles.[59]

The U.S. Women, Peace, and Security Act of 2017 and the *United States Strategy on Women, Peace, and Security* support the Afghan national action plan for implementing UNSCR 1325 and reconstruction efforts by applying a gender perspective to pursue security and peace operations.[60] The RSM has witnessed modest attempts of WPS implementation, as evidenced by the deployment of gender advisors, female engagement and cultural support teams, and U.S. and donor nation gender-specific funding. Despite the call for meaningful inclusion of Afghan women in the security sector in the U.S. WPS strategy and in annual U.S. Afghanistan Security Forces Fund (ASFF) appropriations, RSM leadership has not prioritized WPS efforts or fulfilled the obligations outlined in these authoritative documents. Afghanistan reconstruction efforts have witnessed a modicum of success and the fragile peace talks

---

[57] Charlotte Isaksson, "Integrating Gender Perspectives at NATO: Two Steps Forward, One Step Back," in *Women and Gender Perspectives in the Military*, 229.
[58] "Women, Peace and Security," North Atlantic Treaty Organization.
[59] Belquis Ahmadi and Rafiullah Stanikzai, "Redefining Masculinity in Afghanistan," *United States Institute of Peace Peacebrief* (February 2018).
[60] Egnell, "Gender Perspectives and Military Effectiveness."

exclude Afghan women's meaningful participation in the peace process. Although implementing UNSCR 1325 in Afghanistan is a stated military objective, underinvestment in WPS application at the command leadership level and daily advisor activities remains unprioritized.

Peace Building

More than 20 years after the adoption of UNSCR 1325, women's participation in peace negotiations and presence at the negotiating table remains exceptional rather than the norm. Women's civil society organizations and the international community have acknowledged a women's participation gap in peace negotiations that has garnered much attention in recent years. There is a growing body of literature and emerging consensus on women's participation in peace negotiations and the quality and durability of peace processes. In fact, a study by Jana Krause, Werner Krause, and and Piia Branfors found that between 1990 and 2014, women signed only 13 of 130 peace agreements. Despite the adoption of UNSCR 1325, the number of peace agreements signed by women has not increased.[61] Addressing the women's peace process participation gap is important because post-conflict societies and countries involved in peace negotiations offer a window of opportunity for increasing women's meaningful political participation. The relative absence of women's perspectives in peace negotiations contributes to negative effects on prospects of peace.[62] Krause et al indicates the robust relationship between women signatories and the durability of peace. The linkages established by female signatories and women's civil society groups contribute to increased peace provisions and commitments supporting effective peace arrangements and implementation.

Authors have argued that the role of women in conflict resolution and peace building provides linkages to provisions and access to resources during post-conflict discussions. The post-conflict period provides a window of opportunity for the meaningful inclusion of women to ensure economic, justice, and security institutions are inclusive of the needs of the population. Examples from countries such as Rwanda demonstrate how the inclusion and engagement of women during post-reconstruction can facilitate gender-sensitive programs and resources. In military advising operations, individual advisors serve as a crucial link in the transfer of knowledge and resources to support post-conflict reform. Regardless of the framework for a post-conflict failed state and the capacity-development program implemented, without inclusion of gender perspectives and implementation by well-educated and -trained advisors, capacity-building programs are fruitless. Lack of gender perspectives mainstreamed

---

[61] Jana Krause, Werner Krause, and Piia Branfors, "Women's Participation in Peace Negotiations and the Durability of Peace," *International Interactions* 44, no. 6 (August 2018): 988, https://doi.org/10.1080/03050629.2018.1492386.
[62] "The Heart of the Matter," in *Sex and World Peace*, 103.

into these development programs remains a strategic gap in reconstruction efforts. This study examines these factors and advisor predeployment training and gender implementation on how they affect advising and reconstruction in Afghanistan.

Applying gendered perspectives to conduct building partner capacity and security sector reform with a foreign security force partner involved in active conflict remains a research gap. The literature on applying a gender perspective to military operations remains focused on internal processes versus supporting a foreign military in implementing gender perspectives in military plans, policies, and operations. It is essential to understand the differences between advising the command and advising a foreign security force. These differences are gendered and involved relations between the sexes, power and authority dynamics, and access to resources, to name a few. Understanding these dynamics affects advising strategies and behaviors.

Implementation of gender perspectives in advisor efforts will result in gender-sensitive reconstruction efforts within the ANDSF to meet military outcomes and objectives, creating the conditions of an effective, affordable, and sustainable force.

Chapter Three
Advising in Wartime
A Foreign Policy Tool

Taliban and Afghan Women

In the early 1990s, more than 1,800 girls dressed in skirts and high heels attended Balkh University in Mazar, which was one of the only operational universities in the country.[63] Throughout the twenty-first century, Afghan women's rights had become the basis of political and social reform. The Afghan civil war forced a generation of Afghan males across the Afghan-Pakistan border to grow up in refugee camps in Baluchistan and the Northwest Frontier Province studying in Madrassas. Taught by mullahs, students were taught the basics of Islamic law and memorized the Koran and the prophet Muhammad's sayings. Having no collective memory of their homeland, these Afghan orphans of the war grew up without women and gathered under the all-male brotherhood. Without the company of women, the subjugation of women became the mission of the true believer and a fundamental marker of Taliban governance.[64] As a fanatical band of religious students, the Taliban swept aside the warring factions and put an end to the civil war. Although peace took root, this was a temporary measure leading up to the Taliban's implementation of a draconian regime that banned nearly all social activities, caged women, and took away women's access to education and employment.[65] The closing of girls' schools and banning of women from working outside the home drew women to conduct activities underground and in the shadows of Taliban rule.[66] It is under this context that the social strata of Afghanistan was reformed and led to the conditions for al-Qaeda and other terrorist organizations to take root in Afghanistan. Yet at the same time, it was under these conditions that Afghan women persisted and became resilient for future peace efforts. Under the Taliban, women were regularly stoned to death for trying to flee Afghanistan with non-blood relatives, thus bringing dishonor to their families.[67] The Taliban's declared aims were to restore peace, disarm the population, enforce Sharia law, and defend the integrity and Islamic character of Afghanistan.[68] These aims were at the expense of humane treatment and upholding of Afghan women's rights, and instead subjected women to gross violations of their rights. The Taliban immediately implemented the strictest interpretation of Sharia law ever seen in the Muslim world. Implementation of gender perspectives supports Afghan reconstruction and security efforts by supporting the ANDSF through building partner capacity with an understanding of the gendered root causes of instability and the effects of Taliban rule on the Afghan population.

---

[63] Rashid, *Taliban*, 57.
[64] Rashid, *Taliban*, 32.
[65] Gopal, *No Good Men Among the Living*, 7.
[66] Rashid, *Taliban*, 29.
[67] Rashid, *Taliban*, 5.
[68] Rashid, *Taliban*, 22.

U.S. Response to UNSCR 1325

In response to the UNSCR 1325, the United States adopted its first WPS national action plan (NAP) in 2011.[69] In 2016, the WPS NAP was revised and updated. The WPS NAP served as an initial framework for the United States to outline detailed steps to fulfill the resolution's obligations. The United States took a further step and signed into law the Women Peace and Security Act in October 2017.[70] This law recognizes the critical role women play in promoting more inclusive and democratic societies, essential to countries and regions' long-term stability.

The act called for a U.S. government-wide strategy and further tasked the DOD with developing a specific implementation plan. In July 2019, the *United States Strategy on Women, Peace, and Security* was released and superseded the WPS NAP of 2016.[71] The strategy emphasizes women's meaningful participation in conflict prevention, conflict mediation, conflict resolution, and counterterrorism. The strategy has three strategic objectives that must be achieved, including 1) meaningful participation of women; 2) women and girls are safer, better protected, and have equal access to resources and programs; and 3) institutionalization and capacity of WPS efforts. The strategy's four lines of effort outline actions the U.S. government will take in fulfilling the strategy. In June 2020, as required by law, DOD released the WPS SFIP.[72] The WPS SFIP has three long-term defense objectives:

1. Defense Objective 1. The Department of Defense exemplifies a diverse organization that allows for women's meaningful participation across the development, management, and employment of the Joint Force.
2. Defense Objective 2. Women in partner nations meaningfully participate and serve at all ranks and in all occupations in defense and security sectors.
3. Defense Objective 3. Partner nation defense and security sectors ensure women and girls are safe and secure and that their human rights are protected, especially during conflict and crisis.[73]

As outlined in the Our Secure Future annotated bibliography, much research has discussed the disproportionate effect military conflict has on women and children, highlighting the importance of the WPS agenda in U.S. military operations in Afghanistan.[74] A simple and transparent measure capturing women's autonomy and empowerment toward the goals of women's inclusion, justice, and security, the Georgetown

---

[69] *The United States National Action Plan on Women, Peace and Security* (Washington, DC: Office of the President, White House, 2011)
[70] Women, Peace, and Security Act of 2017, Pub. L. No. 115-68, (2017).
[71] *The United States Strategy on Women, Peace, and Security* (Washington, DC: Office of the President, White House, 2019).
[72] *Women, Peace, and Security Strategic Framework Implementation Plan* (Washington, DC: U.S. Department of Defense, 2020).
[73] *Women, Peace, and Security Strategic Framework Implementation Plan*, 7.
[74] *Just the Facts: A Selected Annotated Bibliography to Support Evidence-Based Policy Making on Women, Peace, and Security* (Broomfield, CO: One Earth Future Foundation, 2019).

Institute for Women, Peace, and Security's WPS index measures 167 countries in these three areas. Afghanistan ranked 166th with an index of 0.373, with Yemen ranking just below it at 167 with a 0.351 index score.[75] All the scales of quantifiable evidence demonstrating the impact of gender inequality and state security reveals a statistically significant relationship with the physical insecurity of Afghan women.[76] Attacks against women and girls was common under Taliban rule.

The Taliban believe that women should be neither seen nor heard because they drive men away from the proscribed Islamic path and spread sexual temptation across Afghanistan.[77] As one example, while Shamisa Husseini was walking with her sister to go to school, two men approached her and asked if she was going to school. They pulled Shamisa's burqa from her head and sprayed her face with acid.[78] Frequent attacks such as this instilled fear in Afghan families who, in attempts to protect their daughters, refused to let them leave the house and go to school. Understanding the content in the WPS index can influence policymakers and military planners to effectively apply programs and resources in the areas outlined to facilitate reconstruction efforts and the necessity of applying a gender analysis and perspective to all military operations.

Resolute Support Mission Afghanistan

Despite personnel and resource constraints, the operational tempo of conventional forces has steadily increased over the past decade. Conventional forces assigned advisor missions, who are educated, trained, and equipped to serve as advisors, are needed to support international security cooperation in a multitude of complex environments ranging from combat operations to reconstruction to peace-building efforts. Effective execution of advising requires a well-educated and -trained and experienced advisor capable of applying gender perspectives in their advisory duties. This topic is essential as the 2018 *National Defense Strategy* affirms allies' and partners' roles to safeguard the free and open international order whose partnerships remain the backbone of global security. The Women, Peace, and Security Act of 2017 and subsequent U.S. government WPS directives call for greater participation of women in peace negotiations and peace building while also encouraging greater gender perspectives in military operations. Through a greater understanding, one may identify the challenges and opportunities for better advisor training and the affects training has on advisor outcomes. Finally, accounting for gendered perspectives in the partner nations, these measures will strengthen the effects of advising.

In December 2001, during the Bonn Conference, it was decided to create a

---

[75] *Inclusion, Justice, Security: Women Peace and Security Index 2019/20* (Washington, DC: Georgetown Institute for Women, Peace and Security, 2020), 2.
[76] "Wings of National and International Relations, Part One: Effecting Positive Change through Top-Down Approaches," in *Sex and World Peace*, 139.
[77] Rashid, *Taliban*, 2.
[78] Rashid, *Taliban*, 21.

NATO-led International Security Assistance Force (ISAF) in Kabul in September 2003.[79] NATO ISAF took the lead for combat-led operations in September 2003 and initiated the command responsibility and obligation to implement UNSCR 1325 into Afghanistan's NATO military operations. From 2003 to the end of 2014, ISAF was focused on U.S. combat-led operations through the ANDSF training.[80] Throughout the ISAF mission's 11 years, it evolved into training and assisting the ASIs and ANDSF. UNSCR 1325 also grew as the UN Security Council adopted subsequent resolutions that highlighted and addressed sexual gender-based violence and called on member states to adopt NAPs and incorporate gender perspectives into military operations.[81] Gendered perspectives at this time were concentrated on the tactical approach, incorporating women into military patrols to serve as searchers and intelligence gatherers.

The ongoing training and advisory effort helped build security capacity and professionalization in the ANDSF, and on 1 January 2015 the ANDSF took the lead for defense and security operations. At the same time, ISAF transitioned to the RSM.[82] In keeping with its commitment to incorporating gender perspectives into NATO mission operations, the implementation of UNSCR 1325 became an RSM NATO military strategic objective. In 2015, Afghanistan adopted its first NAP for implementing UNSCR 1325.[83] To counter the disproportionate effect conflict has on women and girls, implementing gender perspectives in support of the Afghan NAP during military operations must be applied.

As the NATO mission evolved, advising under ISAF evolved from unit-based warfighting functions to the RSM—a noncombat mission focused on functional-based security force assistance train, advise, assist (TAA) mission conducted at the corps and ministerial levels. Functional-based security force assistance is aligned to eight essential functions in which execution of the functions, systems, and processes will help professionalize and yield an affordable, sustainable, and effective ANDSF.[84] Advising remains an essential aspect of the RSM and serves as an entry point for gender perspective implementation.

The United States has two complementary missions in Afghanistan: 1) RSM, conducting functional-based security force assistance; and 2) the U.S.-led counterterrorism mission, Operation Freedom's Sentinel, against al-Qaeda and the Islamic

---

[79] United Nations Security Council, Resolution 1386, Establishment of ISAF, S/RES/1386 (20 December 2001).
[80] "ISAF's mission in Afghanistan (2001–2014) (Archived)," North Atlantic Treaty Organization, accessed 1 November 2020.
[81] "National Level Implementation," Women's International League for Peace and Freedom, accessed 28 November 2020.
[82] "Resolute Support Mission in Afghanistan," North Atlantic Treaty Organization, accessed 1 November 2020.
[83] "National Action Plan Afghanistan," Women's International League for Peace and Freedom, accessed 28 November 2020; and *Afghanistan's National Action Plan on UNSCR 1325, Women, Peace and Security.*
[84] *Enhancing Security and Stability in Afghanistan*, 7.

State-Khorasan (ISIS-K) and other terrorist organizations operating in Afghanistan. As there is no military solution in Afghanistan, the end state has been political reconciliation with the Taliban and ensuring that Afghanistan is not a safe haven for terrorists to conduct attacks on the U.S. homeland. For the past 19 years, the American way of war has militarily overmatched the Taliban and terrorist organizations. As present-day conditions demonstrate, the U.S. military might not have been enough. Implementation of UNSCR 1325 and the inclusion of gender perspectives in RSM operations and U.S.-led peace talks remains elusive. The reconstruction of an inclusive and sustainable approach in Afghanistan relies on Afghan women's meaningful participation in all sectors of society, including civil, defense, and security sectors. There has been much discussion and increased focus on the importance of understanding gender, conflict, war, and intersectionality in these areas. Intersectional feminist theory is one such approach, which has the potential to facilitate understanding of the Afghan operational environment and gender considerations. Advisors understanding how different social identities such as race, ethnicity, and religion can impact an Afghan's experience of their gender serves as the gateway toward more meaningful engagements.[85] Advisors' knowledge in gender perspective and how they engage with their Afghan counterparts will have a more profound and holistic view of the advising operational environment and increased advisor operational effectiveness.

RSM functional-based security force assistance conducts TAA at the strategic level encompassing the Afghan Security Institutions, including the MOD and MOI with assistance to the National Security Council and Ministry of Finance. Advising at the strategic level is referred to as defense institution building or ministerial advising. At the operational and tactical levels of advising, TAA is conducted at the ANDSF corps and police headquarters, including the Afghan National Army, Afghan Air Force, Afghan National Police, and Afghan Special Security Forces. It is considered security force assistance. The main effort of the RSM is the individual advisor, whose role is critical to the success of the entire mission. Combat advising at the corps level and below is fundamentally different than advising occurring at the ministerial level. Combat advising focuses on manning, training, and equipping the ANDSF to build partner security capacity. Advising at the ministerial level focuses on developing strategy, policy, systems, and processes to build institutional, governmental capacity. An example of combat advising at the corps level and below includes training in individual and collective skills, equipment maintenance and readiness, and combat operational planning to defeat terrorist networks and protect the force. Advising at the ministerial level includes advising ministers on strategic level objectives that are forward-focused on developing the ministry and force in future years.

---

[85] Maria Carbin and Sara Edenheim, "The Intersectional Turn in Feminist Theory: A Dream of a Common Language," *European Journal of Women's Studies* 20, no. 3 (August 2013): 233–48, https://doi.org/10.1177/1350506813484723.

Conclusion

The implementation of UNSCR 1325 in RSM is crucial to military operations. In researching the absence of Afghan women in suicide missions in Afghanistan, Matthew P. Dearing found the enduring presence of a strict culture restricts female participation from society and insurgent organizations.[86] The Taliban's strict policies for the interaction of men and women essentially erased Afghan women from partaking in any role in Afghan society, subjugated and confined them to household duties. Limiting female participation has resulted in a generation of Afghans and 50 percent of the population who have been denied basic rights such as education, health, and social skills. It is under this Afghan world view which reconstruction advisory efforts and peace operations are executed by the military application of power. It is therefore necessary to implement gender perspectives in all military operations.

---

[86] Matthew P. Dearing, "Like Red Tulips at Springtime: Understanding the Absence of Female Martyrs in Afghanistan," *Studies in Conflict and Terrorism* 33, no. 12 (November 2010): 1088, https://doi.org/10.1080/1057610X.2010.523861.

Chapter Four
Advising Afghan Security Institutions and Afghan National Defense and Security Forces

Advising Gender Analysis

As outlined in NATO's strategic guidance, a gender analysis requires the systematic gathering and examination of information on gender differences and on social relations between men and women to identify and understand inequities based on gender. Just as this applies to the planning and execution of military combat operations, a gender analysis is also applicable to noncombat advisory functions, specifically the gendered differences between advisors and how they execute advising duties. Understanding advisor gender perspectives provides insights on advising behaviors stemming from different cultural backgrounds, social relations, military experience, professional occupational specialties, and advisory predeployment training. Cumulatively, all these elements directly influence the execution of the RSM advisory mission. Therefore, determining how an advisor's predeployment training and gendered perspective affect the RSM advising mission provides critical insight into an important aspect of advising the ANDSF.

Of the 30 advisors interviewed, 22 advisors were officially designated as advisors predeployment, with the remaining 8 serving as advisors after they arrived in Afghanistan.[87] Although rank varied across the advisors, it is worth highlighting that 12 officers (majors/captains) served as ministerial advisors responsible for advising senior-ranking Afghan officers, sometimes with ranks two to three ranks above the coalition officer. When asked "what are some limitations to your advising mission," rank mismatch was a common response. Rank mismatch was problematic for advisors because their Afghan counterparts questioned the advisor's experience and legitimacy based on rank and perceived age. Supporting field research, a key finding in the 2019 SIGAR report noted staffing field advisors' challenges due to rank and specialty requirements.[88] Staffing challenges, such as position vacancies, shifting priorities mid-deployment, and lack of advising experience, often resulted in advisors being reassigned advising duties, frequently resulting in rank mismatches.

The cultural diversity of advisors selected to participate in interviews represent the true nature of multinational operations in the RSM headquarters. The United States had the most significant contribution of advisors, totaling 19 and encompassing both military and civilian advisors. The average years of military experience were 21 years. It is worth noting that—excluding U.S. advisors—there was a trend for NATO nation

---

[87] To be officially designated as an NATO RSM advisor, the billet had to be coded on the NATO crisis establishment document, which dictated the required predeployment training.
[88] *Divided Responsibility: Lessons from U.S. Security Sector Assistance Efforts in Afghanistan* (Washington, DC: Special Instructor General for Afghanistan Reconstruction, 2018), 36.

advisors to have extended military service, which could positively impact institutional knowledge.[89] For example, two NATO Colonels had 34 and 38 years in the military working alongside U.S. colonels, who had 29 years each.

Age-rank distribution demonstrates different gender perspectives stemming from experience. This age was observed in how Afghans perceived advisors and often reflected in the acceptance of advising support or rejection of advice. Several advisors noted their senior and older Afghan counterparts struggled to get things accomplished because "some Afghan Colonels don't work well . . . because of rank issues." However, as senior Afghan officials were replaced with younger Afghans, rank issues became less of a problem. This behavior change can be explained through RSM efforts to fill vacant Afghan ministerial positions with educated Afghans, who tended to be younger and eager to work in the ministry. There were no significant gender differences between the advisors and Afghan advisee sex. Female coalition advisors were spread across the RSM headquarters organization, providing a wide array of feedback on individual relationships. Citing the most critical factor was relationship building and the establishment of trust over time, female advisors emphasized that frequent face-to-face contact supported advising efforts.[90] Male Afghans were receptive to their female advisor counterparts and appreciated the assistance. Overwhelmingly, when asked "In your opinion, are there any advisor skills that should be taught to advisors? Please describe?" female advisors cited the need for cultural understanding, understanding different personalities, listening to Afghans, supporting their needs in accomplishing the mission, and for interpersonal skills. Differences in sex between advisors and advisees presented no hindrances to good advisor-advisee relationships if the advisor was culturally aware and sensitive to and respecting of Afghan culture. The fact that many Afghan advisees were receptive to female advisors serves as an indicator of the importance of understanding. Afghans are receptive to assistance regardless of coalition advisors' sex and support future efforts to implement UNSCR 1325 in the ANDSF, which calls for gender equality.

As expressed by all advisors, one of the most significant limiting factors of advising was the volatile security situation, which directly impacted escort security requirements for advisors to conduct off-base face-to-face advising.[91] This burden decreased the advisor's available time spent at the Afghan advisee's office and limited access to advising opportunities. As one advisor reported during their interview, not being

---

[89] Exceeding 35 years for ranks of lieutenant colonel and colonel, which in the U.S. military is usually seen in senior colonels and general officers respectively.
[90] Advisors reported they met face to face at least four to five times per week.
[91] The security situation required advisors to have guardian angels provide security during all off-base advising sessions. It required two web-based applications: one for movement request to advising location, which included two vehicles and three guardian angels, and one for approval to conduct the advising mission. This administrative burden limited availability of guardian angel and security support to conduct frequent advising missions. Priority of vehicle and guardian angel support went to general officers and senior ranking colonels serving in tier 1 and 2 advising positions, such as the national security advisor, minister of defense, first deputy minister, and chief of general staff.

colocated with her advisee during the workday hampered their relationship. Furthermore, the Afghan advisee was blacklisted for inappropriate behavior toward the U.S. government, resulting in a lack of advising contact for over a month. Once the U.S. government lifted advising restrictions, the damage was done, as the advisor-advisee relationship was strained. Field research gathered three findings addressing the impact of advisors' predeployment training and gendered perspective on the RSM advising mission.

Finding 1: Senior Leadership Advocacy

RSM is a NATO-led mission with a U.S. Army four-star general serving as its commander and as the U.S. Forces Afghanistan (USFOR-A) commander.[92] In this unique role, the commander has the roles, responsibilities, and authorities of both missions. Under the NATO authority, RSM is mandated by its operational plan to implement UNSCR 1325 as a NATO strategic military objective. The obligation of WPS implementation is codified in U.S. law.[93] Despite these mandates, UNSCR 1325 implementation remains a strategic gap and shortfall. Put bluntly, WPS, gender perspectives, and key leader engagement discussions have not been prioritized. Gender perspectives are excluded mainly from RSM TAA advising activities. Although NATO has embraced gender initiatives and perspectives, U.S. leadership has not, thus impacting critical decisions and advising functions integral to ANDSF TAA operations.

To highlight the importance of senior leadership advocacy, advisors highlighted the influence of the RSM commander's personal communication guidance, as articulated in *The Rules Resolute Support Book* and frequently used buzzwords such as *flat communication, reliable partners, institutional viability,* and *advise at the point of need,* which were frequently cited in advisors interview answers.[94] These buzzwords were often incorporated into advisors' discussions with their Afghan partners, emphasizing group thinking and repetition at the behest of senior leadership. Buzzwords were also broadcasted across advisors' computer screensavers and posted in high-frequency common areas such as the laundry facility, gym, and dining facility. Operationalizing the commander's buzzwords remained elusive, as many advisors lacked predeployment advisory training and the broad concept of security sector assistance in which these terms applied. As a result, buzzwords were repeated time and again and incorporated into advising vocabulary.

Additionally, UNSCR 1325 provides WPS implementation guidance through the NATO/EAPC *Women, Peace and Security Policy and Action Plan 2018, Bi-Strategic Command Directive 040-001 Integrating UNSCR 1325 and Gender Perspective into the NATO Command Structure* (Bi-SCD 040-001), and the *Allied Command Oper-*

---

[92] *Enhancing Security and Stability in Afghanistan,* 10.
[93] National Defense Authorization Act for Fiscal Year 2020, Pub. L. No. 116-92 (2019).
[94] *The Rules Resolute Support Book,* vol. 14 (Commander, Resolute Support, April 2020).

ations (ACO) Gender Functional Planning Guide.[95] The NATO mandate was critical to the execution of the gender advisor mission. As DOD rolls out its WPS SFIP, it has much catching up to do to be on par with NATO RSM partner nations to implement UNSCR 1325 and gender perspectives. As an example, the Bi-SCD 040-001 translates political direction to the implementation of UNSCR 1325 through the integration of gender perspectives and gender mainstreaming in all activities within the strategic commands. To accomplish this task, a gender analysis must be conducted during all stages of planning and execution. According to the Bi-SCD 040-001, "gender analysis requires the systematic gathering and examination of information on gender differences and social relations" within a given area of operations "to identify and understand inequities based on gender."[96] As the U.S.-Taliban peace deal demonstrated and targeted attacks on Afghan women serving as politicians and lawyers and on women in the security sector increased in frequency, it is evident gender perspectives in NATO military operations largely remain excluded, further demonstrating the lack of adherence to its strategic directives. In a mission such as RSM with 38 nations and numerous partner nations, all with different cultural backgrounds, WPS provides a common framework to direct attention, focus, and efforts for reconstruction, stability, and peace. WPS is advocating for gender equality so all Afghan society members can have a say in peace; protection of women, children, and men from the harmful effects of conflict; and the resolution of conflict to achieving long-lasting peace. Champions of the WPS agenda in Afghanistan need to see beyond the physical number of women in uniform. As of April 2020, 5,257 women served in the ANDSF.[97] Meaningful participation means women participate and are able to communicate their thoughts to make a decision, which fosters a sense that their contributions matter and are well represented in the ANDSF.

As an example, advisors who advised Afghan women reported an advising limitation stemmed from their advisee being interrupted or ignored by Afghan male leadership. Furthermore, Afghan women would frequently be excluded from important meetings where decisions were made. The lack of Afghan female participation in decision-making meetings serves as an indicator for coalition advisors to advocate on behalf of their Afghan advisee to be included in decision-making discussions.[98] In order for women to be included, senior leaders must engage their Afghan male coun-

---

[95] *NATO/EAPC Women, Peace and Security Policy and Action Plan 2018* (Mons, Belgium: NATO/Euro-Atlantic Partnership Council, 2018); NATO TSC-GSL-0010/TT-170733/Ser: 0761, *Bi-Strategic Command Directive 040-001: Integrating UNSCR 1325 and Gender Perspective into the NATO Command Structure* (Mons, Belgium: NATO, 17 October 2017); and *Allied Command Operations (ACO) Gender Functional Planning Guide* (Mons, Belgium: Supreme Headquarters Allied Powers Europe, 2015).
[96] *Bi-Strategic Command Directive 040-001 Integrating UNSCR 1325 and Gender Perspective into the NATO Command Structure*, 5.
[97] *Enhancing Security and Stability in Afghanistan*, 46.
[98] Only a few Afghan women served in key positions within the Ministry of Defense headquarters. Their absence from decision-making discussions and meetings was evident. Research did not evaluate if there was a difference in decision-making presence if the advisor was a female and the Afghan a male.

terparts and have these discussions. NATO has reaffirmed their financial commitment and support of the ANDSF until the end of 2024.[99] This commitment underscores the importance of integrating gender perspectives so those gender considerations are seamlessly woven into all future advising practices to ensure a gender-sensitive foundation be provided on which the principles of WPS can flourish.

While there has been much focus on professionalizing the ASI and ANDSF to achieve a political settlement with the Taliban, lack of senior leadership support in implementing WPS undermines future advisory and reconstruction efforts. WPS is not strongly communicated or conveyed in strategic messaging. NATO's recent UNSCR 1325 video makes no reference to Afghanistan gender advising.[100] This further demonstrates at the political and military level all-too-common lip service advocacy, but on the ground, WPS execution and implementation in Afghanistan present an entirely different reality. The absence of the preservation of Afghan women's rights in the U.S.-Taliban peace deal and joint declaration demonstrates how NATO/U.S. military operations have neglected their responsibility in ensuring Afghan women are meaningfully participating in peace negotiations in support of the Afghan NAP for implementing UNSCR 1325.[101] The lack of WPS implementation by the United States has also impacted servicemember gender training and predeployment opportunities.

Finding 2: Advisor WPS Operationalization Training

> From the standpoint of America's national security, the most important assignment in your military career may not necessarily be commanding U.S. soldiers but advising or mentoring the troops of other nations as they battle the forces of terror and instability within their own borders.
>
> ~Robert M. Gates[102]

The advising mission in Afghanistan could not have been better illuminated than during former Defense Secretary Gates's speech at West Point in 2008. At the time, the United States was conducting simultaneous advising missions in Iraq and Afghanistan. Touted by the American people as the forever war or the longest war, the U.S. advisory mission in Afghanistan has been the subject of academic and political discourse since the war began. For most of the conflict in Afghanistan, the United States

---

[99] "NATO Allies and Partners Reaffirm their Commitment of Financial Support for Sustainable Afghan Security Forces," North Atlantic Treaty Organization, accessed 1 November 2020.
[100] "NATO and WPS: How an Unlikely Pair Became Inseparable," North Atlantic Treaty Organization, accessed 1 November 2020.
[101] *Agreement for Bringing Peace to Afghanistan between the Islamic Emirate of Afghanistan Which Is Not Recognized by the United States as a State and Is Known as the Taliban and the United States of America* (Washington, DC: U.S. Department of State, 2020); and *Joint Declaration between the Islamic Republic of Afghanistan and the United States of America for Bringing Peace to Afghanistan* (Washington, DC: U.S. Department of State, 2020).
[102] "Text of Secretary of Defense Robert Gates's Speech at West Point," *Stars and Stripes*, 22 April 2008.

and NATO have deployed advisors as individual augmentees or as part of pickup training teams and assigned them according to the needs of the Joint Force. A report from SIGAR in 2017 reported individual military augmentees frequently received little notification for their deployment, which had a direct effect on their predeployment training. Poor predeployment advisor training resulted in advisors being grossly unprepared for their advising mission, directly affecting their ability to understand the local and cultural dynamics of Afghanistan.[103]

In 2019, SIGAR published another report addressing multiple times the negative impact the lack of advisory predeployment training had on Afghan security sector assistance.[104] Field research confirmed predeployment training remained deficient. To inquire about advisory predeployment training, the following question was asked: Did you receive advisory predeployment training prior to deployment? Data revealed 8 advisors received no predeployment training, and 22 received training. To determine the preparation and quality of advisor training, the following questions were asked: On a scale of 1–10 (10 being the best, one being the least), how well did your training prepare you to be an advisor? and On a scale of 1–10 (10 being the best, one being the least), how was the quality of your predeployment advisory training? As depicted in figures 7.4 and 7.5, the average advisor rating of training preparation was 2.72, and the quality of training was an average of 2.

These low ratings are problematic because the NATO RSM mission is centered on the advisory mission, which is reliant on educated and trained advisors. At the ministerial level, it is problematic when advisors are untrained and expected to carry out strategy when unprepared. An analysis of the gender impacts of untrained advisors results in advisors relying on their personal experience and cultural background to conduct their advising functions. This detracts from command unity of effort and synchronization across the advisory mission. Inadequate predeployment training, advisors' unfamiliarity with the current mission efforts, and limit the advisors' ability to be fully immersed with the commander's priority and mission objectives.

---

[103] *Reconstructing the Afghan National Defense Security Forces: Lessons From the U.S. Experience in Afghanistan* (Washington, DC: Special Instructor General for Afghanistan Reconstruction, 2017), 44.
[104] *Divided Responsibility*, x–xv, 8, 12–16, 21–23, 36–38, 46–55, 58, 62, 143, 151.

200 WOMEN, PEACE, & SECURITY

Figure 7.4. Training Preparation

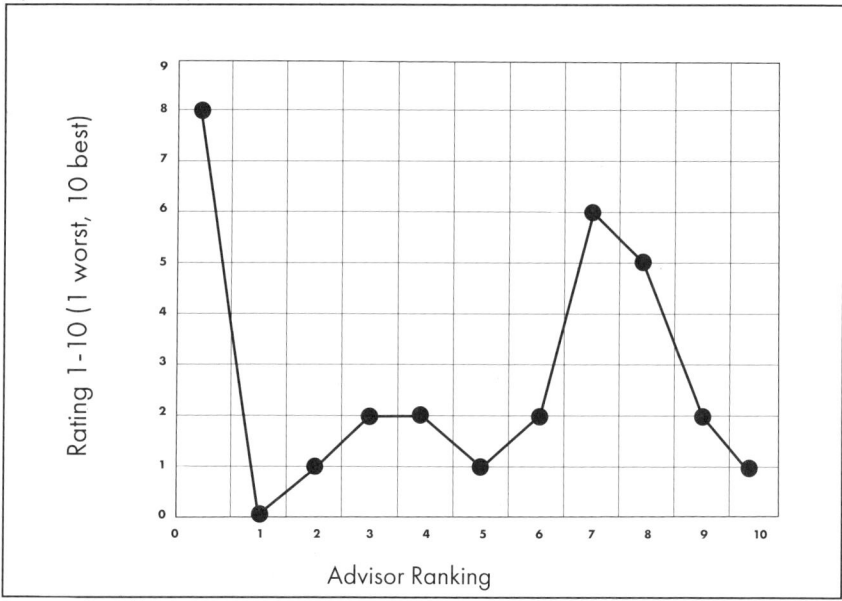

Source: adapted by MCU Press

Figure 7.5. Quality of Training

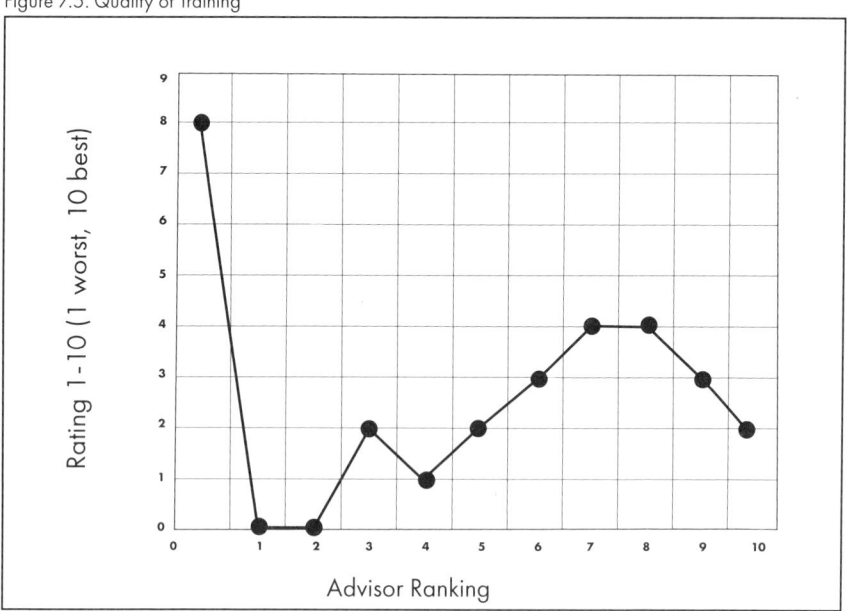

Source: adapted by MCU Press

As a tool to advance the interests of the United States in Afghanistan, the United States employed advisors at the tactical to strategic levels. A SIGAR report in June 2019 discusses field advising and ministerial advising, and the report highlighted that theater-specific training remained a gap.[105] The military advisor is the action arm of the military instrument of national power and the tip of the spear when it comes to executing advising. Advisors must be educated and trained in gender and cultural perspectives with the knowledge, skills, and attributes (KSA) required to support the professional development of the ANDSF. Gendered perspectives will facilitate overcoming existing structural barriers, such as cultural, language, and gender barriers, which are often present in military advising and situations, to foster a sense of neutrality and partnership within the advising context. Gendered perspectives will also facilitate the balancing from that of a strict combat advising mindset to that of a strategic advising mindset. Each has a unique set of skills for execution. This is not to say that an advisor should not have a combat mindset. Rather it is recognition of advising that supports their ANDSF advisee to develop their own KSA that is gender-sensitive to their operations. Advisors are not certified teachers. But they should have access to policies, doctrine, existing academic research, and published articles to know how to train and educate others.

In 2017, during an end-of-deployment interview, Major General Richard Kaiser, then the commanding general of Combined Security Transition Command Afghanistan (CSTC-A), noted how advisor training was lacking, which had a direct impact on the advising mission.[106] Additionally, Major General Willard M. Burleson, the Minister of Defense's deputy advisor, also noted the same observation and the impact the lack of advisor training has on the conduct of the advising mission.[107] It is worth noting that Major General Burleson coauthored the Army's *Multi-Service Tactics, Techniques, and Procedures for Advising Foreign Security Forces*, which was not a predeployment requirement for advisors to read before assuming duties as an advisor.[108] Despite these in-depth interviews identifying gaps in advisor training and acknowledging educational and training advisor gaps, the Army decided on a new advising concept through the employment of security force assistance brigades. In an attempt to "prepare Afghan foreign security forces to secure their nations," this new Army toolkit resorted to solving a problem of its previous attempts to professionalize the ANDSF through untrained advisors.[109] The employment of SFABs in Afghanistan

---

[105] *Divided Responsibility*, 7, 41.
[106] *News from the Front: Ministerial Advisors—Combined Security Transition Command Afghanistan (CSTC-A)* (Fort Leavenworth, KS: Center for Army Lessons Learned, U.S. Army Combined Arms Center, 2017), 2.
[107] *News from the Front: Advising at the Ministerial Level in Afghanistan—Insights from Major General Willard M. Burleson III* (Fort Leavenworth, KS: Center for Army Lessons Learned, U.S. Army Combined Arms Center, 2018), 3.
[108] *Multi-Service Tactics, Techniques, and Procedures (MTTP) for Advising Foreign Security Forces.*
[109] Wesley Morgan, "The Army's Latest Weapon to Turn Around the War in Afghanistan," *Politico*, 26 January 2018.

did not address nor resolve the lack of predeployment advisor training. Instead, it was a futile attempt to provide additional staffing to an advising problem set that was initially designed in an ad-hoc fashion. Advisors deployed to keep up with the constant rotation of forces, often at the expense of receiving adequate advisor training, which after 19 years were tired, exhausted, and worn out.

The rebuilding of the ANDSF is reliant on advisors capable of transferring knowledge to their foreign security force partner. This can only come through advising education sensitive to gender roles and perspectives in Afghan society and how these gender perspectives affect advising at all levels: strategic, operational, and tactical, as highlighted by SIGAR in numerous reports. Advisors who lack sufficient education and training often ask themselves: what is the advisor's mission to train, advise, and assist? Advisors deficient on Afghan cultural education and training lack the knowledge of Afghan social politics, such as the Taliban's draconian rule and trauma inflicted on Afghans, especially Afghan women. The lived experience of an Afghan serving in the ANDSF was shaped by the Taliban's governance and it influences their behavior. After 20 years of coalition intervention and fighting against the Taliban, the inclusion of Afghan gender roles and perspectives into military operations must be considered to continue to professionalize the ANDSF and prevent the ANDSF from reverting to ethnic rivalry and fighting within the security sector. Engagement with the assistant minister of defense for personnel and education (AMOD P&E), the senior-most woman in the Ministry of Defense, provides an example of the necessity of gendered advising perspectives.[110] Serving as a strong source of gender advocacy in the Ministry of Defense, she was seen as the lead advocate of any issue dealing with Afghan women's gender initiatives in the Ministry of Defense and Afghan National Army. RSM general officers and representatives from the international community and Embassy staffs supporting Ministry of Defense and Afghan National Army gender initiatives would meet with AMOD P&E to discuss gender initiative programs. However, engagement with AMOD P&E on gender initiatives was problematic. The lead for gender initiatives in the Ministry of Defense and Afghan National Army is the director for human rights and gender integration (HRGI). Since this was a civilian position rather than a ministerial position, interdepartment tension developed because AMOD P&E took the lead on gender initiatives without including the HRGI director. The tension between both departments grew because advisors lacked gender perspectives and cultural understanding of the role authority influences Afghan engagements. A June 2020 biannual report to Congress noted that "implementation of UNSCR 1325 and NAP 1325 requires senior-level representation and authority requiring promotion of Gender Directorate positions to levels of AMOD."[111] Elevating the HRGI position would resolve authority conflict and account for cultural interaction gendered perspectives.

---

[110] *Enhancing Security and Stability in Afghanistan*, 41.
[111] *Enhancing Security and Stability in Afghanistan*, 41.

It is worth noting that an article published in 2019 highlighted the 1st Security Force Assistance Brigade (SFAB) and assignment of teams across Afghanistan to conduct the Army's re-energized TAA mission. These new SFAB teams, which are permanent, replace previous advising teams, which were ad hoc formations. These 12-person formations are supposed to be jack-of-all-trades not with the goal of teaching the Afghans how to fight, but "to teach them how to sustain the fight."[112] Sustaining the fight requires Afghan soldiers to be literate and capable of synthesizing information and making decisions. As part of the organizational construct, SFABs were educated and trained to perform advisory missions. Afghan gender perspectives remained excluded from predeployment training, incorporating valuable classes on SIGAR lessons learned. In 2017, SIGAR reported the impact of high rates of Afghan illiteracy, an uneducated force, and its advisory mission. The gendered implications of sustained armed conflict in Afghanistan devastated the Afghan educational system, with an estimated 3.7 million children absent from school, with 60 percent of those absentees girls.[113] In an interview discussed in the Afghanistan Papers, as reported in 2016, only about 2 in 10 Afghan military recruits could read and write.[114] Low soldier literacy rates continues to impede security force development as illiteracy is a continued reportable item in semiannual congressional reports. To address the literacy gap, the Ministry of Defense inaugurated the beginning of literacy classes at the Afghan National Army recruitment command.[115] Had gender-sensitive training been conducted to address Afghan soldier illiteracy, SFAB teams would have known that most Afghan soldiers lack basic literacy to read manuals common to vehicles, weapons, and gear. This example is one of hundreds that further highlight the challenges advisors experience by being unfamiliar with the gender perspectives of the Afghan environment.

NATO's support for UNSCR 1325 implementation calls for gender perspectives in military operations.[116] NATO has developed comprehensive gendered training toolkits to be used for deploying servicemembers.[117] Partner nations, such as Canada, have developed a Gender-Based Analysis Plus (GBA+) toolkit, which is used to facilitate a gender-sensitive operational analysis of the mission by identifying how men and women, girls and boys experience the effects of military operations differently.[118] Advisors receiving gender perspectives in their predeployment training would be introduced to these tools to aid them in their advising duties. Despite having gender

---

[112] Sean Kimmons, "Advising at the Corps and Below, US Soldiers Ensure Afghans Are Ready to Fight," Army.mil, 8 October 2019.
[113] "Education: Providing Quality Education for All," UNICEF, accessed 1 March 2020.
[114] "The Afghanistan Papers."
[115] "Literacy Program in the First Time Inaugurated for Afghan National Army," Afghanistan Ministry of Defense, accessed 30 September 2020.
[116] "Gender Perspectives in NATO Armed Forces," North Atlantic Treaty Organization, accessed 1 November 2020.
[117] "HQ SACT Office of the Gender Advisor," North Atlantic Treaty Organization, accessed 1 November 2020.
[118] "What Is Gender-based Analysis Plus," Government of Canada, accessed 1 November 2020.

advisors in the RSM, lack of advisor education, training, and knowledge relating to women or gender is deemed a women's issue. One advisor's gendered actions can be used as an example.

A U.S. advisor was assigned to the Afghan National Army sergeant major. A sexual harassment case at one of the corps had involved an incident with three Afghan female soldiers and an unstated number of Afghan males. The Afghan corps commander reached out to the Afghan sergeant major to address the sexual harassment cases, particularly in handling the Afghan women. The Afghan sergeant major asked his advisor for assistance, who reached out to this author as the Ministry of Defense gender advisor. Upon coordination, the HRGI director reached out to the corps commander to help support the command for investigation and the victims for relief and recovery support. In this particular case, Afghan social roles between men and women influenced Afghan male behavior above his leadership responsibilities and determined that any matter affecting a woman was perceived as a women's issue and automatically referred to by the gender advisors. This gender blindness has resulted in advising problems, as outlined previously and detailed in multiple SIGAR reports.[119] In a recent publication by the Joint Chiefs of Staff discussing the development of today's Joint officers, the publication discusses the need for strategic thinkers and Joint officers who are better educated with foresight in mind.[120] A glance over the document makes no mention of gender perspectives called out in the Women, Peace, and Security Act and the U.S. WPS strategy. This is problematic because senior Service leadership must be the strongest advocates for WPS gender initiatives, which must be articulated in documents emanating from the Joint Chiefs.

Gender perspective training would also provide answers to clarify WPS tenets and to counter negative perceptions of the agenda. WPS is not a women's issue in Afghanistan. It is not a competition between men and women. Instead, it is grounded in collaboration, unity, and interdependence to work toward a common outcome: meaningful participation of Afghan women whose contributions can achieve a long-lasting peace in Afghanistan and end the decades-long armed conflict that has ravaged generations. It is also about inclusion of women as equal participants in conflict resolution and peace building. WPS sheds a new light and way to understand the complex operational environment and seeks to find less-militarized solutions to peace in Afghanistan and to serve vital U.S. interests. Military advisors untrained in gender perspectives is not a sustainable status quo.

When the advising mission comprises well-educated and -trained advisors, the advising mission may see improvements in the outcome of Afghan policies and procedures. Lack of leadership support results in inadequate gender perspective training,

---

[119] *Afghanistan National Defense and Security Forces; Enhancing Security and Stability in Afghanistan;* and *Reconstructing the Afghan National Defense Security Forces.*
[120] *Developing Today's Joint Officers for Tomorrow's Ways of War: The Joint Chiefs of Staff Vision and Guidance for Professional Military Education and Talent Management* (Washington, DC: Joint Chiefs of Staff, 2020).

affecting advisors' knowledge of gender perspectives and the crucial role it plays in the professionalization and development of the ANDSF. Lack of leadership support and gender training leads us to the last critical component of WPS institutionalization.

Finding 3: WPS Institutionalization

Finally, gender mainstreaming in daily plans, policies, and operations is the end state of WPS institutionalization in the RSM and ANDSF operations. States that foster gender perspectives highlight the importance of gender equality to facilitate state security.[121] The reconstruction of Afghanistan has been predicated on a political peace settlement with the Taliban and reduced military operations. For the peace settlement to be effective and sustainable, Afghan women and the preservation of their rights must be centered. Research conducted by the United Nations Women has shown a connection between women's participation and the peace agreement quality.[122] Another study shows that women's contributions to peace processes make peace 35 percent more likely to last for 15 years. Women's participation provides greater diversity and thought into peace building and peacemaking as women have different security needs and promote priorities that challenge dominant state security narratives.[123] With the current RSM efforts of securing a peace deal, gender institutionalization could not come at a more crucial time. The only pathway forward is an Afghan solution that involves Afghan women.

Advisors were asked if they had read the Resolute Support Operational Plan (RS OPLAN). Knowledge of the RS OPLAN would have exposed advisors to the RSM mission, two NATO military strategic objectives, including implementation of UNSCR 1325. Other keywords such as functional-based security force assistance, security force assistance, and defense institution building are also included in the RS OPLAN. Nineteen advisors did not read the OPLAN. This is problematic because 12 of the respondents had never been to Afghanistan. Furthermore, only 20 advisors received a turnover, and many felt their turnovers were limited in scope and needed more turnover to adequately learn their job. These statistics demonstrate the criticality of WPS institutionalization as it provides a framework that develops a common advising culture and understanding the RSM objectives to support advisors' engagements. The WPS agenda transcends political objectives and is critical for achieving U.S. national objectives, vital interests, and the alliance's common defense. There is an inherent tension between the political nature of the WPS agenda and security sector reform and principled military action rooted in gender equality, gender inclusivity, and gender diversity. Military leaders need to understand these lines of effort are complementary and must not be coopted by political and social change merely to satisfy objectives

---

[121] "Roots of National and International Relations," 3.
[122] *Women's Participation in Peace Negotiations: Connections Between Presence and Influence* (New York: United Nations Women, 2012).
[123] Marie O'Reilly, Andrea Ó Súilleabháin, and Thania Paffenholz, *Reimagining Peacemaking: Women's Role in Peace Process* (New York: International Peace Institute, 2015).

that are not aligned with maintaining military readiness and effectiveness. WPS in Afghanistan must be implemented by senior leadership, advisors, and civilian leadership. WPS must be institutionalized and framed within a security lens to achieve long-lasting effects. Extensive guidance on NATO and partner nations on the military contribution to the implementation of UNSCR 1325 largely remains excluded from senior leadership advocacy and WPS institutionalization. This lack of RSM institutionalization affects Afghanistan. As an example, First Vice President Amarullah Saleh stated on 3 November 2020 that the Kabul University attack was a failure of intelligence—the same intelligence that RSM coalition advisors are supporting.[124] As another example, it could be reasonably assumed the Kabul Maternity clinic attack in May 2020 could have been prevented through better gendered intel analysis. The attack on Fawiza Koofi in August 2020 followed by a September 2020 U.S. embassy in Afghanistan warning about an increased security threat and targeted attacks against women.[125] Gendered intelligence analysis identifies vulnerable Afghans and facilities needing extra precaution during a sensitive transitional phase in Afghan politics and the peace-making process.

WPS reframes the concept of warfighting and military objectives, offering a gendered mission analysis observation: the Afghan center of gravity is women and children. The Taliban's worldview on the role of Afghan women and seclusion from society offers the coalition an opportunity to influence future action. Engaging women, who represent 50 percent of the population, is essential to countering the Taliban and its radical narrative of Islam. WPS institutionalization is central to supporting Afghan women and girls to achieve future security and stability through their meaningful participation in all aspects of Afghan society. The ongoing political process negotiations and participation of Afghan women within the peace process remains crucial. As has been repeatedly stated, there is no military solution to the conflict in Afghanistan. Therefore, the only solution is a civil solution reliant on the meaningful inclusion of Afghan women.

Gender institutionalization is a critical component of Afghan counterterrorism operations. The utilization of female engagement teams (FETs) and present-day cultural support teams (CSTs) demonstrate the active implementation of gender in the terrorist fight.[126] Female-only teams allow access to an otherwise excluded population while addressing the gendered root causes of conflict. WPS institutionalization approach to professionalizing the ANDSF supports not only Afghan women's specific initiatives, but it will also support challenges such as pay issues, facility issues, and food issues that plague the ANDSF. Though military leadership recognizes the benefit of institutionalization as it applies to the ANDSF at all levels, increased professionalization, effective and sustainable defense, and security institutions will be realized. The coali-

---

[124] Ahmad Mukhtar, "Attack on Kabul University in Afghanistan's Capital Leaves at Least 19 Dead," CBS News, 2 November 2020.
[125] Kathy Gannon, "US Embassy in Kabul Warns of Extremist Attacks Against Women," AP News, 18 September 2020.
[126] Egnell, "Gender Perspectives and Military Effectiveness."

tion welcomes the benefits that derive from WPS leadership support, gender training, and WPS institutionalization, and we are compelled, on the other hand, to think about their implications on peace, security, and stability.

Advising the ANDSF requires senior leadership advocacy, advisor training, and institutionalization to effectively implement UNSCR 1325 and gender perspectives. Support of reconstruction and peace operations requires knowledge on Afghan cultural and social norms in order to effectively engage Afghans in the professionalization of the ANDSF. As advocates, senior leaders are responsible for implementing UNSCR 1325 in support of the Afghan NAP. Advisors lacking Afghan cultural education and training inhibit the effectiveness of professionalization of the ANDSF by not factoring in cultural sensitivities, which could inhibit progress. Lastly, institutionalization of UNSCR 1325 and gender perspectives is imperative for the future stability and negotiated peace settlement with the Taliban.

Chapter Five
Recommendations and Conclusions

Recommendations

Research on the Challenges to Implementing UNSCR 1325. More research is recommended to understand advisory challenges to implement UNSCR 1325. A more in-depth research study on coalition perceptions will provide a holistic view on the unique challenges of WPS implementation across the entire RSM and across all command structures. In-person interviews are recommended to gather personal insight and the human emotions and behaviors expressed when discussing the challenges of implementing a rather straightforward policy. Surveys in this particular case are not adequate as they often miss out on opportunities for advisors to explain aspects of advising challenges. Advising is a personal interaction between two individuals who are influenced by partnerships of other advisors and their personal engagements. When taken cumulatively, advising patterns and behaviors begin to emerge that can indicate gaps and deficiencies as well as strengths and weaknesses to support future advising training and education. In-person interviews are more likely to capture the human aspect of advising and provide greater insight into advising practices. Additionally, research on ANDSF challenges to implementing the Afghan NAP from the Afghan perspective is warranted. Afghan feedback will provide information to better program resources to address gender challenges and address advisory gaps in gender-sensitive advisory actions.

Mandatory Gender-Sensitive Advisor Predeployment Training. Effective advising is reliant on the knowledge, skills and attributes of the individual advisor. Predeployment training should not be compromised or waived when staffing requirements dictate the filling of a vacant position. Advisors are the action arm of Afghan foreign policy implementation and is where milestones succeed or fail. Advisor predeployment training informs advisors on the commander's intent, mission set, and country-specific advising skills. Training also affords the opportunity to synchronize individual advising efforts, achieving unity of effort in advising duties. Annual training monitoring and evaluation feedback from advisors provides relevant feedback to ensure training is relevant and updated with the current operational environment. Despite SIGAR advisory lack of predeployment training lessons learned, the DOD continues to send advisors to the RSM inadequately trained. The authors personal experience deploying as a gender advisor without adequate training validates this deficiency. Predeployment training was focused on combat skills and tactical level of advising, which was not adequate preparation for the ministerial level of advising. The billet to which this author was assigned was designated as requiring the completion of the NATO Gender Advisor Course. Due to DOD force generation process procedures, the deployment timeline did not offer ample time to attend the course, and the author was deployed without the training. This resulted in crucial time spent learning the nuances of gender advising on the job, rather than learning the billet and duties of a gender advisor, wasting significant personnel hours. Advisors in this same situation cost the RSM significant lost time in conducting the advising mission.

WPS Implementation. The Office of the Secretary of Defense must be more involved in the implementation of UNSCR 1325 in the RSM and Operation Freedom's Sentinel missions. Interest driven from the top down signals the importance of implementing 1325 to achieve DOD objectives. WPS implementation must have senior leader buy-in and be incorporated into all Office of Secretary of Defense-RSM briefings and discussions. The disconnect between DOD's WPS SFIP and RSM implementation underscores the importance for OSD level oversight.

Shortly after arriving in Afghanistan, the author was contacted by OSD and Central Command (CENTCOM) to discuss gender funds that were identified to spend in Afghanistan, presenting an opportunity to leverage the U.S. authority and gender initiatives support to Afghanistan to discuss concerns about achieving the objectives as outlined in the Women, Peace, and Security Act and the U.S. WPS strategy. At the time the WPS SFIP had not been released. After an exchange of emails back and forth, no further guidance or support was provided in addressing the challenges the author was having in implementing gender perspectives into the RSM. Seeking further guidance, the author reached out to the RSM higher headquarters gender advisors to seek clarification and guidance. Under the NATO authority, RSM reports to the Allied Joint Force Command Brunssum (JFCBS), which reports to Supreme Headquarters Allied Powers Europe. Each of these commands has a gender advisor who, unfortunately, provided little to no assistance in supporting the execution and implementation of UNSCR 1325. This was due in part to a new gender advisor in the JFCBS command who had little knowledge on the gender advising mission and was preoccupied with advising the JFCBS command on gender implementation and had no capacity to support the RSM advising mission. Strategic-level focus was also observed with the NATO gender advisor. This resulted in missed opportunities for higher headquarters to assess, monitor, and evaluate the efficacy of strategic-level guidance and policies in implementing UNSCR 1325 in Afghanistan.

Gender-Sensitive Budgets. Support of Afghan women in the ANDSF requires a commitment of funding to support building partner capacity programs unique to Afghan women. Execution of funding requires gender-sensitive budget execution protocols to ensure funding is not lost in the approval process. A recommendation to achieve a gender-sensitive budget is the establishment of a parallel gender project approval process to allow gender project approval in a timely manner. Approved gender projects will enhance women's participation in the ANDSF to increase capacity and meet the objectives in the Afghan National Action Plan 1325.

During the author's discussion with OSD and CENTCOM, officials offered operations and maintenance gender funds and inquired if the money could be spent. Feedback was twofold. 1) access to funding was not an issue. The fiscal year 2020

National Defense Authorization Act appropriated up to $45.5 million to support Afghan gender programs and activities.[127] 2) Gender money was not an issue, rather it was the execution, as in the spending of money, that was challenging. Despite lengthy discussion into the specific challenges of spending Afghan gender initiatives money, no action was taken on the part of OSD nor CENTCOM to rectify this issue. As a result, in June 2020, for the second year in a row, the Afghan Ministry of Defense did not execute any gender funds through its own fiscal budgeting process. Gender proposals and the associated budget packages were not signed off by Ministry of Defense officials, thereby missing deadlines by which to execute gender money. The effect this had on the Afghan National Army was that additional capacity building programs such as education, language training, and female empowerment training opportunities went unfunded.

Conclusion

Through their advising support of the ANDSF, NATO and U.S. military leadership remain at the forefront of Afghan reconstruction and peace negotiations. While the current withdrawal plans of U.S. and coalition forces is ongoing, advising support and sustained gender-sensitive funding are imperative to ensure Afghans can negotiate a peaceful outcome to the conflict and increase gender equality. Military leadership needs to embrace a new approach in Afghanistan and must be comfortable talking about gender and the initiatives outlined in support of the Afghan NAP and WPS agenda. Even in conservative societies such as Afghanistan's, gender perspectives are already being discussed in forums such as civilian causalities, human rights violations in the ANDSF, sexual misconduct, and meaningful participation of women in the ANDSF. What remains to be seen is how gender perspectives change these discussion outcomes to enable an effective institution that meets national obligations of humanitarian and civil laws, which is the core of the WPS agenda.

Implementation of UNSCR 1325 and the Afghan NAP through leadership support will continue to be challenged by the dominant male patriarchal culture in Afghanistan and the reluctance of U.S. military leadership. What has been demonstrated in Afghanistan is that there is no cookie-cutter approach to addressing Afghan society's challenges and gender issues. Afghan women must be empowered to be change-makers and they need support to allow WPS principles to take root and flourish. One thing is clear, the lack of gender perspectives before, during, and after conflict will continue to have a disproportionate effect on women and girls if WPS principles are not incorporated into military planning and operations. The United States must confront what the future of Afghanistan will look like using a gendered lens; talk about the uncomfortable issues, set aside egos, and come to terms that military action alone cannot subdue terrorists. The United States is at a critical juncture in Afghanistan. To maintain

---

[127] For more information, see National Defense Authorization Act for Fiscal Year 2020, Pub. L. No. 116-92 (2019).

the gains achieved over the past 19 years, the United States must fully embrace WPS as the mechanism for true peace and stability in Afghanistan so it will not serve as a terrorist haven from which to launch attacks on the homeland.

President Joseph R. Biden's announcement of U.S. troop withdrawal from Afghanistan by 11 September 2021 is four months past the 1 May deadline as agreed to in the U.S.-Taliban peace deal. In an interview, Afghanistan ambassador to the United States Roya Rahmani stated that "the ball is in the Taliban's court."[128] For the past 20 years, the Taliban has justified its war on the basis of the presence of foreign troops in Afghanistan. As troops withdraw to meet the 11 September deadline, the Taliban has no justification for continuing its war on Afghan citizens. The role of Afghan women in society is directly tied to the Afghan government's advocacy to foster meaningful participation and inclusion of women in the peace process, decision-making bodies, and the security sector. The rights of Afghan women are directly impacted by the security situation, and should security deteriorate, women and girls will experience disproportionate impacts as they once did under the Taliban governance in the 1990s. U.S. lawmakers express grave concerns for Afghan women in ceding control to the Taliban, whose draconian governance curtailed women's rights.[129] Billions of dollars in support of Afghanistan's reconstruction has significantly improved women's and girls' access to education, public life, and health services. Yet despite promises of having a seat at the table, Afghan women continue to be underrepresented in peace talks led by Ambassador Khalilzad, who has not advocated or made it women's inclusion a priority.

Since the signing of the U.S.-Taliban peace deal in February 2020, Afghan women remain targets of violence; several prominent women activists have been shot, killed, and assassinated.[130] If the international community, including the United States, truly believes in the enormous amount of research conducted on the successful role of women in peace processes, then—in accordance with the U.S. Women, Peace, and Security Act and the WPS SFIP—DOD and DOS must aggressively advocate for the increased active participation of Afghan women in the Afghan government and the peace process. Leveraging the Women, Peace and Security Act of 2017, the president of the United States must verbally advocate and support Afghan women to take the lead for the peaceful reconstruction of their country. The future of Afghanistan is on the shoulders of Afghan women and girls, who need reassurance in their cause for a peaceful Afghanistan. In support of this effort, the RSM should transfer the gender advising mission to the Security Cooperation Office at the U.S. Kabul Embassy. Within this office, the establishment of a Women, Peace and Security

---

[128] Jack Detsch, "Afghanistan Ambassador: 'The Ball is in the Taliban's Court'," *Foreign Policy*, 23 April 2021.
[129] Patricia Zengerle and Jonathan Landay, "As U.S. Troops Leave Afghanistan, Lawmakers Fear Dark Future for Women," *Reuters*, 27 April 2021.
[130] "Bureau of Educational and Cultural Affairs, Honorary Posthumous IWOC Group Award (Afghanistan)," U.S. Department of State, accessed 30 April 2021.

section would continue ANDSF gender advocacy and be aligned to continue supporting gender initiatives in coordination with the international community through the U.S. Agency for International Development. Preventing Afghanistan from becoming a terrorist safe haven will not happen overnight, nor will it occur with the absence of Afghan women's participation in society. Despite the trauma Afghan women have suffered under the Taliban, they are willing to negotiate and talk with the Taliban. The United States and the international community must give Afghan women the chance.

PART 8

Implementing Women, Peace, and Security

Operationalizing Women, Peace, and Security in the Armed Services
Army Strategic Implementation Plan

by Major Danielle Villanueva, U.S. Army[*]

Chapter One
Introduction

This study explores how the Women, Peace, and Security (WPS) agenda is operationalized in the U.S. armed forces with specific emphasis on the Army. An analysis of publications and archival documents, interviews with subject matter experts, and a case study of WPS implementation in the Australian Defence Force helped inform this effort to operationalize WPS and identify best practices and to introduce a WPS Army strategic implementation plan (WPS ASIP). This plan is an opportunity to provide a critical enabler for the emerging global peace and security context and fulfill the Service's legal requirements under WPS.

In 2001, predicated on the role of women in conflict prevention and resolution, the UN passed UNSCR 1325, which called for the full participation of women in peace and security initiatives. UNSCR 1325 and the eight subsequent resolutions provide the "international framework for the implementation of gender perspective in the pursuit of international security and the conduct of peace operations."[1] UNSCR 1325 coined the term Women, Peace, and Security to encompass a broad array of topics specifically related to the impact of armed conflict on women and girls and the importance of their contributions to conflict resolution and peace building. The follow-up resolutions included a range of complex, multilayered issues such as the representation of women in conflict resolution; gender perspectives mainstreaming; training reformations; and protection of women, girls, and boys from conflict-related threats. Eventually, the United States began its own body of founding documents, including the most recent DOD *Women, Peace, and Security Strategic Framework and Implementation Plan* released in June 2020. The plan provided three main defense objectives, which included to model and employ WPS within our own formations, to promote women's participation for partner nations, and to promote the protection of partner nation civilians. The Army is the foremost land service branch of the United States and the largest component of DOD. As of now, there is not a comprehensive plan for how the Army will operationalize WPS. As we enter a period of complex,

---

[*] The views expressed in this chapter are solely those of the author. They do not necessarily reflect the opinion of Marine Corps University, the U.S. Marine Corps, the U.S. Navy, the U.S. Army, U.S. Army War College, the U.S. Air Force, or the U.S. government.
[1] Robert Egnell, "Gender Perspectives and Military Effectiveness: Implementing UNSCR 1325 and the National Action Plan on Women, Peace and Security," *Prism* 6, no. 1 (March 2016): 73.

multidomain conflict, the armed forces, specifically the Army, must capitalize on every opportunity to build capabilities and increase security.

At the Service level, efforts to incorporate WPS are ongoing and mostly focused on professional military education and incorporation into doctrine. A U.S. Army WPS strategic implementation plan is necessary to synchronize efforts across the Service to better guide tactical, operational, and strategic decisionmaking and warfighting. The recommendations given in this paper focus on mainstreaming a gender perspective and seek to bridge the gap between policy and operationalizing WPS. The current evolving nature of war and the threats facing the United States demand a greater emphasis on all warfighting tools beyond hard security tactics and strategies.

Chapter Two
Literature Review

The study of gender and security largely began with the post–Cold War reevaluation of international relations theory.[2] Over the course of the next three decades, the field of gender studies has expanded to include positive benefits within the security sector, from military and peacekeeping effectiveness to broad security outcomes.

The relationship between gender and the military evolved through a gradual, albeit swift, progression of scholarship beginning with feminist international relations theories and empirical approaches to women's participation. The scholarship further evolved into a small body of work that evaluates how women improve the effectiveness of military and peacekeeping organizations. Finally, more recent scholarship explores the relationship between broader security outcomes and military actions.

Feminist International Relations Theory

The root of WPS lies in feminist international relations, which emerged in the 1980s. Cynthia Enloe's *Bananas, Beaches and Bases* (1989) began a series of intellectual studies focused on how the international system relies on masculinity and femininity and the often-overlooked work of women.[3] J. Anne Tickner's paper "Man, the State, and War: A Gendered Perspective on National Security" emphasizes the importance of considering war and conflict through a gendered lens accounting for the experience of all people, specifically women.[4] The scholarly work of Jean Elshtain explores the different roles of women in war from "beautiful souls" or innocent noncombatants to their service as soldiers and how these gendered dimensions shape politics and problem solving as a state.[5] Collectively, feminist international relations theory seeks to illustrate that women and gender construct a clearer picture of international politics and, subsequently, peace, war, and conflict.

Quantitative Analysis

Moving to a more quantitative analysis, scholarship explores empirical data about women's participation and outcomes in the field of international relations and security. In Valerie Hudson's article "What Sex Means for World Peace," she emphasizes that the situation and security of women in a country is often the best indicator of

---

[2] Marysia Zalewski, "Feminist Approaches to International Relations Theory in the Post–Cold War Period," in *The Age of Perplexity: Rethinking the World We Knew* (New York: Penguin Random House, 2018).
[3] Cynthia Enloe, *Bananas, Beaches and Bases: Making Feminist Sense of International Politics* (Berkeley: University of California Press, 2014), 125–73.
[4] J. Ann Tickner, "Man, the State, and War: A Gendered Perspective on National Security," in *Essential Readings in World Politics*, ed. Karen A. Mingst and Jack L. Snyder (New York: W. W. Norton, 2004), 94–100.
[5] Jean Bethke Elshtain, *Women and War* (New York: Basic Books, 1987).

how likely that country is to be involved in conflict.[6] Her empirical results lead to the conclusion that human security (namely the security of women) is linked to national and international security.[7] The scholarly work of Mary Caprioli evaluates gender equality and state aggression, providing analytical data linking the degree of gender equality and women's role in the state to the likelihood of the state to use force during an interstate dispute.[8] This body of scholarship introduces the idea and provides analytical data that meaningful participation of women and gender equality can have further implications for conflict.

Operational Effectiveness

Beyond feminist theory and quantitative analyses, much of the scholarly writings and research on WPS focus on how its principles increase operational effectiveness and unit functionality. A number of works in this vein look at women's involvement in UN peacekeeping missions, for example. The UN Department of Peacekeeping Operations (DPKO) was the first military organization to consider gender perspectives.[9] DPKO, partnered with the UN Division for the Advancement of Women, conducted a comprehensive study of peacekeeping operations in Bosnia, Cambodia, El Salvador, Namibia, and South Africa. The study demonstrated that women on peacekeeping teams improved access and support for local women, made men more reflective and accountable, increased capability, and decreased conflict and confrontation.[10] Subsequent studies have shown that when 30 percent of mission personnel are female, local women more quickly join the peace effort, increasing the effectiveness of peace agreements and leading to better stability of the state. A stable state is less likely to harbor terrorists, violate human rights, and require intervention from the international community.

Scholars supporting gender inclusion within the armed forces similarly conclude an increased credibility. One such scholar, Sahana Dharmapuri refers to increased credibility as providing a greater opportunity to build trust and mitigate violence among the local population.[11] A well-known example is the all-female police units from India deployed in a peacekeeping capacity in Liberia. These women police are seen as more approachable and make the key victims of conflict-related violence

---

[6] Valerie Hudson, "What Sex Means for World Peace," *Foreign Policy*, 24 April 2012.
[7] Hudson, "What Sex Means for World Peace."
[8] Mary Caprioli, "Gender Equality and State Aggression: The Impact of Domestic Gender Equality on State First Use of Force," *International Interactions* 29, no 3 (October 2003): 195–214, https://doi.org/10.1080/03050620304595.
[9] Sabrina Karim, "Women in UN Peacekeeping Operations," in *Women and Gender Perspectives in the Military: An International Comparison*, ed. Robert Egnell and Mayesha Alam (Washington, DC: Georgetown University Press, 2019), 23. The DPKO underwent a name change effective 1 January 2019 and is now Department of Peace Operations (DPO).
[10] Donna Bridges and Debbie Horsfall, "Increasing Operational Effectiveness in UN Peacekeeping Toward a Gender Balanced Force," *Armed Forces and Society* 36, no. 1 (October 2009): 125, https://doi.org/10.1177/0095327X08327818.
[11] Sahana Dharmapuri, "Just Add Women and Stir," *Parameters* 41, no. 1 (Spring 2011): 60.

feel safer.[12] In addition, the presence of women peacekeepers deterred sexual and gender-based violence and was viewed as more attuned to the needs of the local populations.[13]

Furthermore, scholars cite the creation of female engagement teams (FET) and cultural support teams (CST) in Afghanistan as an example of increased unit effectiveness through enhanced information gathering.[14] FETs, CSTs, and similar programs were used to engage and search a previously underutilized portion of the population, developing a better understanding of local conditions and increasing force protection of troops in the area of operations.[15] Admiral William McCraven, U.S. Navy (Ret), noted that the inclusion of CSTs enabled greater access and action to the local population, boosting traditional military information support as well as medical and civil affairs activities contributing to mission effectiveness.[16]

When discussing operational effectiveness and gender, peace, and security, there is danger in marginalizing women to stereotypical roles, proliferating the idea that only females performing in these roles can contribute to mission success. For example, there are essentializing assumptions that women peacekeepers or those in the armed forces are inherently best placed to gather information from or protect female civilians.[17] These assumptions risk limiting the potential for meaningful contribution and do not increase women's participation "beyond gender stereotypes and 'add women and stir' calls for parity."[18]

Societal Outcomes and Military Actions

The last area of scholarship (and the most recent to emerge) examines the relationship between military action and broad security outcomes. In "Through a Gender Lens: The Need for Robust Research into Diversity and Military Effectiveness," Lieutenant Colonel Jeannette Haynie, argues that leaders must use every tool at their disposal to inform a clearer picture of security and develop assumptions. She argues that well-developed and effective tactical, operational, and strategic plans must incorporate diverse perspectives, specifically a gender lens, at every level of leadership. Finally, Haynie suggests that gender is still largely dismissed as irrelevant to "real"

---

[12] *Gender Sensitive Police Reform in Post Conflict Societies* (New York: United National Development Programme, 2007).
[13] *Gender Sensitive Police Reform in Post Conflict Societies.*
[14] *Gender Sensitive Police Reform in Post Conflict Societies.*
[15] Robert Egnell and Mayesha Alam, "Introduction: Gender and Women in the Military—Setting the Stage," in *Women and Gender Perspectives in the Military*, loc. 296 of 6776, Kindle.
[16] William H. McRaven, "Women in Special Operations Forces: Advancing Peace and Security through Broader Cultural Knowledge," in *Women on the Frontlines of Peace and Security* (Washington, DC: National Defense University Press, 2014), 127–28.
[17] Gretchen Baldwin, "Expanding Gendered Understandings Key to Protection Concerns," IPI Global Observatory, 15 November 2019.
[18] Baldwin, "Expanding Gendered Understandings Key to Protection Concerns."

security, ignoring the established links between diversity and outcomes.[19]

Along the same lines, in "A Cornerstone of Peace: Women in Afghanistan," the authors argue that the military must fully embrace and capitalize on its internal diversity to effectively engage with partner nations and leaders at all levels must fully understand "the linkages between the security of women and the security of the state."[20] The piece connects the full implementation and integration of WPS in the security sector, particularly the armed forces, with meaningful security assistance as an essential component for U.S. success in future conflict.

Consistent throughout the scholarly work on women, peace, and security is the argument that women and gender belong in and enhance the study of security. The field of gender and security has rapidly evolved within the last 30 years, and the divisions in the literature between theory, quantitative analysis, organizational effectiveness, and broader security implications have and will continue to evolve as studies expand. The division in literature is directly influencing and informing divisions on where and how policy and implementation of WPS is applied and integrated. However, there is a significant gap in literature from the implementation at a policy level to integration into military operations, which this paper explores further.

---

[19] LtCol Jeannette Gaudry Haynie, USMCR, "Through a Gender Lens: The Need for Robust Research into Diversity and Military Effectiveness," *Women Around the World* (blog), Council on Foreign Relations, 29 October 2019.

[20] Kyleanne Hunter, Jeannette Gaudry Haynie, and Natalie Trogus, "A Cornerstone of Peace: Women in Afghanistan," *War Room* (blog), U.S. Army War College, 8 January 2021.

Chapter Three
Background of WPS in the United States

Following the publication of international policy and coinciding with the evolution of WPS scholarship, the United States developed state-level policy and guiding documents. In 2011, President Barrack Obama signed Executive Order 13,595 establishing the *United States National Action Plan on Women, Peace, and Security*. In 2017, the Women, Peace, and Security Act was signed into law, strengthening efforts for the meaningful participation of women in conflict prevention and peace building. The law ensures congressional oversight of how the United States promotes and implements women's meaningful participation in conflict prevention and resolution. The United States released a national strategy on WPS in 2019 outlining four primary lines of effort.

1. Seek and support the preparation and meaningful participation of women in conflict related decisionmaking.
2. Promote the safety and protection of women's and girls' human rights.
3. Adjust U.S. international programs to improved outcomes in equality for, and the empowerment of, women.
4. Encourage partner nations to adopt policies to improve the meaningful participation of women.[21]

To achieve the goals outlined across the four lines of effort, the DOD released the *Women, Peace, and Security Strategic Framework and Implementations Plan* (WPS SFIP) in June 2020. The SFIP organizes WPS implementation along three defense objectives that include modeling and employing WPS, promoting partner-nation women's participation, and promoting the protection of partner nation civilians. The WPS SFIP further dissects each objective and provides intended effects.[22] Following the release of the WPS SFIP, DOD issued a memorandum outlining the guidance for implementation that included a series of data calls to document progress. The data call requires DOD entities to report on a series of indicators supporting the defense objectives outlined in the WPS SFIP. The indicators include the number of high-level commitments on WPS led by DOD, the funding expended in support of WPS objectives, the number of public statements by high-level officials on WPS, the number of doctrine changes to support WPS, and the number of training curricula that integrate WPS. The memorandum calls for DOD entities to include lessons learned to further refine metrics and best practices on operationalizing WPS.[23]

---

[21] *United States Strategy on Women on Women, Peace, and Security* (Washington, DC: Office of the President, White House, 2019), 6.
[22] *Women, Peace, and Security Strategic Framework and Implementation Plan* (Washington, DC: U.S. Department of Defense, 2020).
[23] Under Secretary of Defense, memo, "Implementation of the Department's Women, Peace, and Security Strategic Framework and Implementation Plan," 23 October 2020.

To date, none of the Services have implemented collective, systematic plans; most of the services and their affiliated professional military education (PME) institutions are implementing individual WPS strategies, led largely by individual change agents in leadership positions. For example, the U.S. Army War College recently signed a charter on WPS, which officially seeks to integrate WPS principles into its curriculum.[24] The Marine Corps University Command and Staff College spearheaded a WPS community of interest, which also recently passed a charter and is currently exploring curriculum modifications to include gender analysis and perspectives.[25] The U.S. Naval War College recently created the position of WPS chair to better assimilate WPS topics into PME and coordinate among different communities of interest.[26] Though each of the Services has taken initiatives to meet the SFIP objectives, the actions vary and lack standardization.

In order to analyze how military Services operationalize WPS, a study of existing data in publications and archival data was conducted. Additionally, seven semistructured interviews—three gender advisors, one gender focal point, and three cultural support team members—were conducted to help inform a comprehensive WPS implementation plan. Interview participants were selected based on their background and experiences as gender advisors or cultural support team members. The interviews were 30–40 minutes long and were recorded via note-taking. Interview participants were found through contacts at the UN DPKO. Finally, a case study of WPS implementation within the Australian Defence Force using publications and two semistructured interviews help inform WPS best practices.

Operationalizing WPS in the Armed Forces

This study uses the defense objectives outlined in the 2020 WPS SFIP as a framework to discuss current WPS implementation efforts in the U.S. armed forces. It is important to note that since the armed forces do not currently have a codified WPS program, much of the information has been provided through a series of semistructured interviews with subject matter experts and through analysis of existing data.

Defense Objective 1

Defense Objective 1 states, "The Department of Defense exemplifies a diverse organization that allows for women's meaningful participation across the development, management, and employment of the Joint Force." The WPS SFIP goes on to specify that the DOD should model and implement the WPS principles it encourages in part-

---

[24] U.S. Army War College Public Affairs Office, "USAWC Signs Women, Peace, and Security Charter," press release, 27 October 2020; and news item, Office of the Provost, Director of Women, Peace, and Security Studies, Women, Peace, and Security, U.S. Army War College, accessed 13 May 2020.
[25] J. M. Bargeron, president, Marine Corps University, memo, "Marine Corps University Women, Peace and Security Committee Charter," 14 January 2021.
[26] U.S. Naval War College, "U.S. Naval War College Seeks Woman, Peace, and Security (WPS) Chair Associate/Full Professor AD-1701-05/07," press release, 20 December 2020.

ner nations and to continue to model and advocate for meaningful participation of women. Across the Services, different initiatives are underway or have been started to support Defense Objective 1.[27]

To address the modeling portion of Defense Objective 1, the armed forces have focused on increasing the number and capacity of women within the ranks. A majority of these initiatives focus on what this paper will refer to as *structural barriers*—items or systems that inhibit career progression or lead to decreased retention of women. The most notable initiative is the 2015 lifting of the ban on women in combat and the integration of women into those previously closed combat arms billets. Additionally, the Services have taken a number of administrative measures that consider the recruitment and retention of women such as primary and secondary caregiver leave, enhancing deferred deployment options for birth mothers, modifying grooming and hairstyle policies, and reevaluating child-care options in light of the COVID-19 pandemic. The 2020 Defense Advisory Committee on Women in the Services (DACOWITS) annual report provides a comprehensive summary of recommendations on matters and policies relating to the recruitment of servicewomen in the U.S. armed forces and is used to inform policy changes that the Services have made in the past.[28] While recognizing that continued analysis and revision of policy related to structural barriers is essential to the meaningful participation of women in the armed forces, the research in this paper does not address structural barriers in recommendations for WPS implementation, but recommends further research on the subject.

Another way the armed forces have implemented Defense Objective 1 is through the use of gender advisors and gender focal points. The combatant commands have championed the use of gender advisors to incorporate gender perspectives and human security considerations into campaign plans, operations, and training. Generally speaking, combatant commands attempt to follow NATO *Bi-Strategic Command Directive 040-001* guidance as it pertains to gender advisors and gender focal points. However, the commands lack internally published guidance or explanation of the structure and training associated with these initiatives. In the absence of a codified gender advisor or focal point construct, a series of interviews with current and former gender advisors and gender focal points helped paint a clearer picture of roles and responsibilities, current structure, and training requirements. Gender advisors (GENADs) are personnel whose sole responsibility is to provide guidance to commanders on how to incorporate a gender perspective into operations and missions. A gender focal point (GFP) is often located in subordinate units or staffs and supports the GENAD in operationalizing gender perspectives. The role of a GFP is usually secondary to the primary role the individual has within their respective unit or staff sections. The location

---

[27] *Women, Peace, and Security Strategic Framework and Implementation Plan*, 7.
[28] *Defense Advisory Committee on Women in the Services 2020 Annual Report* (Washington, DC: Department of Defense Advisory Committee on Women in the Services, 2021).

of the GENAD on the staff varies between different combatant commands, with some located in the Operations Section (J3), Strategic Planning and Policy (J5), or the Civil Military Cooperation section (J9). Three out of four of the GENADs interviewed stated that the GENAD should have a place on the special staff with direct report authority to the commander. Additionally, former and current GENADs stated that there should be GFPs within the J3, J5, and J9, as well as staff synchronization functions to foster persistent coordination.[29]

The training associated with the GENADs and GFPs varied among the individuals interviewed. Two personnel interviewed had completed a gender operationalization course offered by a combatant command and two individuals had not received any training due to cancellations as a result of the COVID-19 restrictions. The U.S. Army's Peacekeeping and Stability Operations Institute website provides information on a Joint-Certified Operational Gender Advisor Course to train personnel. The article implies that the course was rotating through the combatant commands with a future plan of residing at one location within the United States; however, research did not discover any updated information past December 2018.

Additionally, Joint Knowledge Online offers training modules on integrating gender perspectives into operations and on the role of GENADs.[30] The specific training and training level expected of the combatant command GENADs and GFPs varied greatly between organizations and lacked codified prerequisites and requirements. In the December 2019 DOD WPS overview brief, none of the armed forces indicated integrating a formal GENAD or GFP program. In the same brief, the Army specified that it would "provide subject matter expertise on WPS principles such as gender integrations, female engagement teams, and gender perspective within Army component support to CCMD theater security cooperation."[31]

Integrating WPS into various Service-level PME is another way the Services implement Defense Objective 1. Specifically, the Naval War College and the Naval Postgraduate School seek opportunities to incorporate WPS into their curricula and activities. The Marine Corps Command and Staff College offered a "Gender, War, and Security" elective in the 2020–21 academic year and is exploring options to incorporate WPS initiatives into exercise planning, wargaming, and the core curriculum. As they have not incorporated WPS across all curricula and activities, at this time, PME institutions continue to develop their integration. In addition to inclusion in PME, the Services complete annual training requirements for sexual assault awareness and

---

[29] Gender advisor, interview with author, 6 January 2021; gender advisor, interview with author, 29 January 2021; gender advisor, interview with author, 8 February 2021; and gender advisor, interview with author, 11 February 2021.
[30] "Joint-Certified Operational Gender Advisor Course," Peacekeeping and Stability Operations Institute, 23 July 2020.
[31] "Department of Defense Women, Peace, and Security" (PowerPoint presentation, Joint Certified Operational Gender Advisor Course, U.S. Army Peacekeeping and Stability Operations Institute, Carlisle, PA, December 2019), hereafter WPS PowerPoint presentation.

combatting trafficking in persons with additional training given to deploying troops. However, there are not courses available that are specific to WPS.

The same progress can be seen on including WPS pillars in training exercises at both the Joint and Service level. The Army has expressed efforts to incorporate WPS into combat training center rotations. WPS was incorporated into a U.S. and Australian Joint exercise called Talisman Sabre in 2015, which marked the first appearance of a WPS component in a large-scale Joint training exercise.[32] In order to mainstream a gender perspective into all levels of planning, the armed forces must include WPS training objectives in exercises and activities.

During the December 2019 DOD WPS overview brief, the combatant commands stated their intentions to reference WPS in their respective theater campaign plans.[33] Several Joint publications mention gender and women, peace, and security, such as *Joint Personnel Support*, Joint Publication (JP) 1-0, which includes a section on WPS, or *Joint Planning*, JP 5-0, which includes gender considerations and highlights the necessity for a gender advisor. The Army is currently updating regulations such as pamphlets and doctrinal manuals as they come up for revision. *Stability*, Army Doctrinal Publication (ADP) 3-07, includes a section on WPS that focuses on incorporating objectives from the 2016 WPS national action plan (NAP) where appropriate.[34] *Protection of Civilians*, Army Techniques Publication (ATP) 3-07.6, and the *Protection of Civilians Military Reference Guide*, second edition, emphasize gender perspectives and sexual and gender-based violence.[35] References to WPS have recently been included, or are in the process of being included, in policy and doctrine at both the Joint and Service levels. However, most of the doctrine mentioned deals with stability operations or protection of civilians. In order for WPS to be effectively implemented and considered, considerations must be included in the deliberate planning process doctrine and Service-level guiding documents.

Defense Objective 2

Defense Objective 2 states, "Women in partner nations meaningfully participate and serve at all ranks and in all occupations in defense and security sectors." The WPS SFIP affirms that the United States will adjust security cooperation programs and work with allies and partners to promote inclusion of women at all levels of defense and

---

[32] Susan Hutchinson, "Leading the Operationalization of WPS," *Security Challenges* 14, no. 2 (2018): 124–43.
[33] WPS PowerPoint presentation.
[34] *United States National Action Plan on Women, Peace, and Security* (Washington, DC: Office of the President, White House, 2016).
[35] *Joint Personnel Support*, JP 1-0 (Washington, DC: Joint Chiefs of Staff, 2020); *Joint Plannning*, JP 5-0 (Washington, DC: Joint Chiefs of Staff, 2020); *Stability*, ADP 3-07 (Washington, DC: Headquarters, Department of the Army, 2019); *Protection of Civilians*, ATP 3-07.6 (Washington, DC: Headquarters, Department of the Army, 2015); and *Protection of Civilians Military Reference Guide*, 2d ed. (Carlisle, PA: Peacekeeping and Stability Operations Institute, U.S. Army War College, 2018).

security.[36] Prior to the WPS SFIP release, combatant commands were already considering WPS in their operations.[37] In U.S. Southern Command, leaders emphasize women's participation in the security sector during key leader engagement with strategic partners across South America. In 2018, U.S. Indo-Pacific Command started a women's mentorship program to share knowledge and empower women in the Mongolian defense and security sector to build capacity and conduct gender analyses in disaster response efforts.[38]

At the Service level, specifically the Army, an example of promoting women in partner nations is the often-cited use of CST members to train the Afghan Female Tactical Platoon (FTP) supporting the Afghan Special Security Forces (Ktah Khas). An interview with a recent CST member involved in training the FTPs uncovered anecdotal information based on the person's experience.[39] The interviewee observed dwindling support among Afghan and U.S. leadership, stating that leadership did not observe training unless there were dignitaries or political personnel visiting. The interviewee also noted resourcing issues, highlighting funding disparities between the FTPs and the Ktah Khas. The establishment of the FTP is one example of many that demonstrates a clear focus of the U.S. military to promote gender equality and participation of women in partner nation security forces. However, the dwindling support from leaders at all levels, especially as the United States looks to leave Afghanistan after current and ongoing Taliban negotiations, highlights gaps and seams in tactical, operational, and strategic level thinking on why WPS matters for security and seemingly treating it as a neglected collateral duty.

Another example of a potential opportunity to implement and integrate WPS was the Army's development of the Security Forces Assistance Brigade in 2018 to specialize in train, advise, and assist missions. Members received specialized training needed to advise partner nations.[40] While this type of unit seems to be an ideal organization to support defense objective two of the SFIP, a member of 2d Security Forces Assistance Brigade that participated in the 2019 deployment in support of Combined Security Transition Command–Afghanistan confirmed that the brigade's training did not include the 2016 WPS NAP or the 2017 Women, Peace, and Security Act. Additionally, the brigade member, a tactical level leader, did not have any involvement with women in the Afghan defense or security sector and noted that gender perspectives were not a consideration when conducting tactical-level planning with Afghan counterparts. While this is the experience of one individual, the 2d Security Forces Assistance Bri-

---

[36] *Women, Peace, and Security Strategic Framework and Implementation Plan*, 1–22.
[37] Jim Garamone, "DOD Leaders Brief Women, Peace, Security, Program to Congressional Caucus," U.S. Department of Defense, 14 December 2020.
[38] Jim Garamone, "Women, Security, Peace Initiative Military Effective," U.S. Department of Defense, 5 November 2020.
[39] Cultural Support Team member, interview with author, 4 February 2021.
[40] "Army Creates Security Force Assistance Brigade and Military Advisor Training Academy at Fort Benning," Army.mil, 16 February 2017.

gade is the Army's key unit to support the development of a partner nation's military. By not training 2d Security Forces Assistance Brigade members on WPS pillars, the ability to build a partner nation fully committed to WPS is severely crippled. While the armed forces, specifically the Army, have made significant efforts to support and encourage the participation of women in the defense and security sector of partner nations, there are many opportunities that can provide more meaningful and comprehensive security assistance for partner nations.

Defense Objective 3

Defense Objective 3 states, "Partner nation defense and security sectors ensure women and girls are safe and secure and that their human rights are protected, especially during conflict and crisis."[41] The WPS SFIP explains that the department will work closely with partner nations' security sectors to facilitate their ability to ensure the safety of their civilians, especially women and girls. As part of their WPS initiatives, the combatant commands have supported defense objective three in various ways. In the December 2019 DOD WPS overview brief, U.S. Africa Command (USAFRICOM) pledged to execute capacity building with military legal professionals on sexual and gender-based violence and human rights and integrate WPS principles into exercises with partner nations. In its premier annual training event, Flintlock 2019, USAFRICOM integrated WPS themes throughout the exercise to promote meaningful participation and to enhance the ability of key partner nations to provide security to their people, especially women and children.[42]

As previously referenced, all U.S. military servicemembers conduct annual training on combating trafficking in persons as an online course that provides awareness on sexual and labor trafficking scenarios.[43] While this course does not provide in-depth information on sexual and gender-based violence as it pertains to conflict, the course does raise awareness of associated issues such as human trafficking. However, *Stability*, ADP 3-07, has information about including gender perspectives and highlights war crimes affecting women as a special consideration in the "Protection of Civilians" and "Women, Peace, and Security" sections.[44] During the 2019 WPS overview brief, the Army pledged to include sex-disaggregated data and gender-specific data and analysis into the Army Threat Integration Center products.[45] Currently, the military has emphasized defense objective three in a limited capacity through data collection, doctrine, and training. Planning doctrine is a useful tool to help include and plan for gender considerations in operations. The Army could capitalize and include more broad gender considerations with regards to gender-based sexual violence in plan-

---

[41] *Women, Peace, and Security Strategic Framework and Implementation Plan*, 7.
[42] "Women, Peace, and Security," U.S. Africa Command, accessed 9 November 2021.
[43] "Combatting Trafficking in Persons," U.S. Department of Defense, 9 November 2021.
[44] *Stability*, 3-9, 3-11.
[45] WPS PowerPoint presentation.

ning doctrine at the Service and Joint levels and include gender dimensions in training operations. In addition, the Army could expand its annual online course to include protection aspects.

The combatant commands have used gender advisors, gender focal points, education and training, and policy and doctrine to incorporate WPS and implement the DOD WPS SFIP with varying levels of success. Subject matter experts at the combatant commands state that leader buy-in is the number one factor that determines how gender perspectives are integrated into strategic- and operational-level planning and execution. At the Service level, efforts to incorporate WPS are ongoing and largely focused on PME and incorporation into doctrine. Most operational and tactical leaders have no knowledge of the 2016 WPS NAP, the Women, Peace, and Security Act, or how to incorporate gender perspectives into operations. A U.S. Army WPS implementation plan would help synchronize efforts across the Service, thereby enhancing operations and mission effectiveness.

Chapter Four
WPS Implementation
The Australian Defence Force

In researching how other UN member states have implemented WPS in the defense sector, Australia was referenced in scholarly articles as a positive example. Australia's report *Local Action, Global Impact: Defence Implementation of Women, Peace and Security 2012–2018* was published in 2018, giving a few years of practice to analyze and provide lessons learned.[46]

In 2012, Australia launched its national action plan (NAP) on WPS establishing a whole of government approach.[47] The NAP specified 24 actions for the Australian government with the Department of Defence having a role in 17 of the actions. The Australian WPS defence implementation plan is coordinated and presented through six lines of effort: 1) Policy and Doctrine, 2) Training, 3) Personnel, 4) Mission Readiness, 5) International Engagement, and 6) Governance and Reporting.[48] This case study will briefly summarize each line of effort and report progress.

In terms of policy and doctrine (line of effort one), the Australian Department of Defence has made significant progress updating all key strategic guidance documents with WPS operational guidance. This includes but is not limited to the *Defence Corporate Plan*, the Defence Business Plan, Defence Planning Guidance, and Australia's Military Strategy 2016. Additionally, Australia has developed operational directives and orders that include a multitude of WPS considerations for current and future operations. Finally, the Australian Defence Force has developed and updated doctrine in support of integrating gender perspectives. An example of new doctrine is the Australian Defence Force Joint Doctrine Note on Gender in Military Operations.[49] In addition to doctrine at the Joint level, the Royal Australian Air Force's *Gender in Air Operations*, Doctrine Note Series AFDN 1-18, provides a holistic approach and consideration to gender in enhancing air mission success. Interviews with former Australian Defence Force GENADs and members of the Royal Australian Air Force provided insight into gender in operations. Subjects expressed that they were better able to incorporate gender perspectives into operations than their army counterparts due to the air force doctrine helping integrate gender into the planning process. Australia's emphasis on policy and doctrine as part of the gender mainstreaming process is commendable; however, much of the changes are at the Department of Defence level with little updates done at the Service level with the exception of the air force.[50]

---

[46] *Local Action, Global Impact: Defence Implementation of Women, Peace and Security 2012–2018* (Canberra, ACT: Australian Department of Defence, 2018).
[47] *Local Action, Global Impact*, 6.
[48] *Local Action, Global Impact*, 18.
[49] *Local Action, Global Impact*, 26–29.
[50] *Local Action, Global Impact*, 31.

Line of effort two is focused on WPS training of defense personnel. This training includes individual and collective levels. At the individual level, WPS and gender analysis are taught at the Australian Defence College's Australian Command and Staff and Centre for Defence Strategic Studies Courses.[51] The WPS agenda and gender concepts are included in predeployment training and are taught by an experienced GENAD. Furthermore, interviewees stated that the air force developed an online course that is mandatory for all members, which exposes the workplace to gender perspectives in operations. At the collective level, WPS practical scenarios are included in military rehearsal exercises. An example is the incorporation of WPS objectives as a critical part of Talisman Sabre 15, a biannual bilateral military exercise with U.S. counterparts. UNSCR 1325 was referenced in the training objectives and the scenario included gender-based issues. The exercise personnel and staff received pertinent WPS training and integrated core concepts of WPS into their planning.[52] During the exercise, 12 GENADs provided recommendations and consultation.[53] Additionally, specialized training for Defence GENADs and GFPs are required to operate in those positions, while the Australian Defence Force operates its own course to provide required training. As of 2018, Australia had 53 women and 48 men trained as GENADs. While Australia has made concerted efforts to incorporate WPS principles into education, it is still missing from most Service- and entry-level training and education.[54]

The third line of effort focuses on the GENAD and GFP framework and gender balancing efforts. The Australian Department of Defence has established 10 GENAD positions at tactical, operational, and strategic levels to advise (commanders) on gender perspectives. GFPs perform their role as an additional duty and are responsible for integrating WPS principles into their assigned units; however, it is not clear the levels to which these personnel are trained or implemented.[55] The implementation plan mentions working toward increasing the number of women, especially at senior levels, and details that one way they are doing so is removing all gender restrictions. Australia's advisor and focal point structure provides a way to normalize gender mainstreaming. However, the structure may serve as a limitation to the proliferation of gender perspectives and gender mainstreaming. The defence implementation plan does not mention structural or institutional changes associated with gender balancing such as recruiting efforts, retention, parental leave, or child care.

Line of effort four is mission readiness, which seeks to integrate WPS considerations and gender perspectives into the operational planning process.[56] The imple-

---

[51] Local Action, Global Impact, 34.
[52] Hutchinson, "Leading the Operationalisation of WPS," 124–43.
[53] Susan Harris Rimmer, "The Case of Australia: From Culture Reforms to a Culture of Rights," in Women and Gender Perspectives in the Military, 197.
[54] Local Action, Global Impact, 38.
[55] Local Action, Global Impact, 42.
[56] Local Action, Global Impact, 54.

mentation plan highlights a series of exercises and operations that integrated WPS. The fifth line of effort looks externally at international engagement on WPS issues with partner nations through seminars, joint training, shared education, and support of diverse infrastructure aimed at building partner capacity. Finally, line of effort six emphasizes the importance of reporting and governance. This last line of effort, however, does not provide measures of performance and measures of effectiveness to enhance reporting.

The Australian Strategic Policy Institute reviewed Australia's implementation of WPS in a recent special report and noted that Australia has shown leadership in advancing the WPS agenda. The report goes on to say that while Australia has made significant advances in the implementation of WPS, significant inconsistencies and resourcing gaps are still prevalent.[57] The author argues that gender perspectives do not inform Australia's response to international crises, which undermines conflict prevention and stability. The general conclusions offered from two former Australian Defence Force GENADs on advancing gender perspectives in operations were focused on incorporating gender and WPS principles into planning doctrine to ensure a gendered analysis, including sex-disaggregated data, is embedded into all aspects of the planning process, especially wargaming. Interviewees also expressed shortfalls in measures of success and measures of performance due to effects being intangible and requiring leaders to be able to articulate how WPS affects the outcome of operations.

The Australian Defence Force WPS implementation plan provides insight and helps inform an implementation plan in the U.S. military. The Australian Defence Force's 2014 defence implementation plan provides eight years of lessons learned and exposes areas of greater emphasis and improvement.

---

[57] Louise Allen, *Australia's Implementation of Women, Peace and Security: Promoting Regional Stability* (Barton, ACT: Australian Strategic Policy Institute, 2020), 6.

Chapter Five
U.S. Army WPS Implementation Plan
Recommended Framework

There are two main approaches to operationalizing WPS. The first is gender balancing, which refers to equal representation of men and women and equitable distribution of resources and opportunities. For example, the lifting of restrictions to women in combat roles can be seen as gender balancing. The second approach refers to gender mainstreaming or the process of integrating and assessing gender implications of tactical, strategic, and operational level mission planning and execution. It is considering women's and men's interests and varying experiences in planning, policy, programs, and assessments at all levels. The two approaches to operationalizing WPS are largely informed by the theoretical underpinnings split between the overall increased participation and the role of women in the security sector and the broader security outcomes influenced by gender considerations. The following recommendations are organized along four lines of effort focused on gender mainstreaming and will seek to inform operationalization of WPS in the U.S. Army.

Line of effort 1: Seek and support the meaningful participation of women in the military decision-making process and across the development, management, and employment of the U.S. Army forces.
- End state: Women's meaningful participation in and the incorporation of gender perspectives in the military decision-making process will increase and contribute to the U.S. Army's mission effectiveness.

Recommended Planned Actions:
- Develop GENAD and GFP billets at the U.S. Army strategic, operational, and tactical commands in order to better facilitate integrating a gendered perspective into operations. The Army should leverage the training developed by the combatant commands to develop an online training module for GENADs and GFPs that is comprehensive, accessible, and standardized. GENAD and GFP prerequisites and training requirements should be clearly identified.
- Incorporate WPS pillars, gendered perspective, and gendered analysis into PME at all levels. (See sample framework in appendix C, page 249.)
- Include WPS objectives as part of the combat training center rotations training exercises.
- Mainstream the WPS agenda into Army strategic- and operational-level policy.

- Include WPS agenda in Army doctrine as the publications are updated. Develop an Army doctrinal publication specifically addressing gender in army operations. (See sidebar, page 235.)
- U.S. Army personnel preparing for deployment will receive additional instruction on UNSCR 1325, requirements under the Women, Peace, and Security Act of 2017, and gender perspectives in military operations. (See sidebar, page 236.)

Line of effort 2: Address security-related barriers to the protection of human rights of vulnerable populations, safety from violence, abuse, and exploitation, and access to humanitarian assistance.
- End state: Vulnerable populations, including but not limited to women, girls, and boys, are protected from violence, abuse, and exploitation and have better access to humanitarian aid.

Recommended Planned Actions:
- Continue to promote and maintain a zero-tolerance policy toward sexual misconduct through the Sexual Harassment Assault Response Prevention (SHARP) Program. Leaders at all levels remain committed to maintaining an environment of respect for human dignity and free of sexual misconduct.
- Promote and consider respect for gender equality, human rights, and the rule of law through all aspects of military operations and through civil-military cooperation.
- Modify and expand the current Combating Trafficking in Persons training curriculum to include WPS principles as related to protection of human rights with specific emphasis on gender-based sexual violence. The annual requirement should include prevention, indicators and warnings, and appropriate responses for uniformed military personnel.
- Predeployment training for Army servicemembers will include additional instruction on gender-based sexual violence and common security issues and considerations to provide protection to and mitigate risk for vulnerable populations.
- Ensure instruction on security-related considerations to protect vulnerable populations and respond to sexual and gender-based violence is considered during advising and assisting operations.
- At the strategic, operational, and tactical levels, encourage the promotion of women's involvement and leadership in the prevention, management, and resolution of conflict through engagement with local and international government organizations, the UN, and multilateral security forces.

Line of effort 3: Adjust Army internal programs to improve outcomes in women's equality and empowerment.
- End state: WPS agenda and a gender-inclusive approach to conflict resolution are mainstreamed across the Army strategy, capability, and budget planning.

Recommended Actions:
- Incorporate WPS strategy mandate and goals outlined in the 2020 WPS SFIP and apply a gender analysis in the development of future policies, programs, and actions.
- Establish a U.S. Army WPS program coordinator and an Army WPS core working group responsible for coordinating overall implementation of the Army's strategic implementation plan. The group will facilitate learning and best practices on WPS inside and outside of the Service.
- Review and strengthen WPS integration in Army planning, programming, budgeting, and execution.
- Develop and strengthen Army training and resources on WPS concepts, themes, and objectives.
- Encourage senior Army leaders to support high-level engagement on gendered perspectives and WPS-related concepts during strategic-level coordination.
- Provide support and encourage participation in WPS focused seminars, conferences, and working groups.
- Develop, strengthen, and better promote the Army mentorship program to include resources and training that encourage leaders to mentor beyond the chain of command and beyond gender similarities.

Line of effort 4: Encourage partner nations to promote and increase WPS related matters in the international security arena.
- End state: Targeted partner nations make measurable progress to incorporate WPS-related policies and practices that improve the security environment of women and promote the meaningful participation of women in the security sector.

Recommended Actions:
- Leverage bilateral and multilateral opportunities to enhance and integrate WPS such as exercises, operations, and training.
- Apply WPS considerations in providing security force assistance. Seek opportunities to meaningfully engage in the recruitment and retention of women in the defense sector and with women in their security environment to include conflict prevention and resolution and violence against vulnerable populations.

- Support the development and implementation of WPS policies at the strategic, operational, and tactical levels of partner nations' defense sectors.

Sample Outline for Gender in Army Operations Army Doctrinal Publication

I. Gender in Military Operations
  a. Introduction
  b. Background
  c. Definitions and Context
  d. Operational Planning and Execution Considerations
     i. Staff Planning Considerations
  e. Gender Analysis
     i. Introduction
     ii. How
     iii. Documentation and Application
  f. Reporting Requirements and Legal Obligations
  g. Key Principles
  h. Roles and Responsibilities
     i. Commander
     ii. Operations Officer
     iii. Gender Advisor

II. Annexes
  a. Gender analysis example and considerations
  b. Recommended Measure of Effectiveness/Measure of Performance

Recommended Additional Education Requirements for Deploying and Advising Personnel

The Joint Knowledge Online (JKO) course Improving Operational Effectiveness by Integrating Gender Perspective (J3TA-MN1292) is designed to provide an introduction to integrating gender perspectives in military operations and is recommended for all deploying personnel.

The JKO course Gender Perspective (J3OP-MN900-03-11) provides an understanding of UNSCR 1325 and provides ways to incorporate gender issues in advising operations. This course is recommended for personnel deploying in a security force assistance capacity.

Chapter Six
Conclusion and Research Recommendations

WPS policy and implementation has fluctuated between a focus on gender balancing and mainstreaming a gender perspective. The recommendations given in this paper focus on the latter and seek to bridge the gap between policy and actual integration of WPS principles into U.S. Army operations. The current evolving nature of war, expanding a wide diversity of conflict ranging from conventional war to urban terrorism and insurgency, demand a greater emphasis on operationalizing and implementing WPS. Gender relations have a profound impact on state security and conflict and a gendered perspective greatly contributes to the examination and understanding of all aspects of a society and further influences the aims of military operations. Gender considerations further clarify the existing threats and violence the military will have to address in and beyond those presented in traditional warfare. Operationalizing WPS is not a silver bullet, but it can contribute to the military's support of a whole-of-government approach to the United States's far-reaching political goals of democratization, stabilization, economic growth, and proliferation of respect for human rights and rule of law.[58]

Research Recommendations

Structural barriers that hinder the meaningful participation of women in the U.S. Army are a critical focus area for future research related to WPS strategy. During research collection, the cultural issue of military masculinity was often cited as a large barrier to gender mainstreaming. Further research on aspects of military culture that inhibit equity and inclusion and ways to mitigate this is recommended. Lastly, further development of the assessing, monitoring, and evaluation process is necessary to assess progress of the proposed WPS Army strategic implementation plan.

---

[58] Robert Egnell, "Gender Perspectives and Fighting," *Parameters* 43, no. 2 (Summer 2013): 41.

EPILOGUE

The Value of Rewarding Good Writing

It is unlikely we would be reading a monograph like this had it not been for Neyla Arnas's years of dedication to WPS at the National Defense University (NDU). Before her retirement, she shared her ideas (including the NDU WPS Writing Award) with Marine Corps University (MCU) faculty during the 2018 summer Faculty Development Program, and with generous support from the National Naval Officers Association (NNOA), the WPS Writing Award was launched in 2018 as a means of attracting and rewarding WPS-related research and writing among MCU students. The inclusion of this new writing award alongside the other long-standing writing awards continues to serve several important functions. First, the announcement of the award through each college's distance learning platform and an email to all students and faculty informs the MCU learning community about what WPS is and why it is important, and incentivizes student participation through a monetary award. Additionally, the call for papers offers a tangible display for faculty after a WPS class session or panel to help move the yardstick from "So what?" to "Now what?" and give students an opportunity to address WPS-related topics for a research paper with the potential for recognition and reward.

The MCU WPS Writing Award

The MCU WPS Writing Award seeks to draw on the perspective and recommendations of the nation's best and brightest to find unique ways to better leverage the power of all our citizens in conflict prevention and resolution. Papers are judged on their connection to the U.S. WPS Strategy, the United Nations WPS agenda, or the value of a gendered perspective relating to conflict prevention and resolution more broadly. Winning papers emphasize one or more of the following issues:
- Women and conflict resolution
- Women's roles in conflict prevention
- Protections for women during and after conflict (relief and recovery)
- Mitigation of the impact of armed conflict on women and girls
- Comments on the U.S. WPS Strategy
- National and international stakeholders' respective roles in setting and advancing the WPS agenda
- Gender perspectives: The importance of the diversification of thought and viewpoints when planning and executing conflict prevention and resolution

The WPS Writing Award selection committee consists of five members of the MCU faculty who judge the degree to which each submission 1) connects to at least one of the WPS issues as described in the call for papers (listed above); 2) contains a well-thought-out and solidly researched argument; and 3) reflects graduate-level writing. The winner is awarded a certificate of achievement and monetary prize (ranging from $250–$500) from NNOA, presented by the MCU president and NNOA Quantico Chapter president. In years past, the MCU WPS Writing Award winners were:

2019: Major Emily Barton, USMC, Command and Staff College, for her paper titled "Birthing a Cultural Change: Pregnancy and Marine Corps Policy."

2020: Sergeant Dakota Kotula, USMC, College of Distance Education and Training, for his paper titled "Gender-Mainstreaming: Changing Perceptions of Female Engagement in Terror Analysis."

2021: Major Danielle Villanueva, USA, Command and Staff College, for her paper titled "Operationalizing Women, Peace, and Security in the Armed Services: An Army Strategic Implementation Plan"

The "Best of" WPS Writing Award

2021 was the first year that the Joint PME "Best of" WPS Writing Award competition was launched with the goal of bringing together the top papers selected from across the various PME schoolhouses for review by a panel of subject matter experts. The idea emerged as a result of a discussion led by Colonel Veronica Oswald-Hrutkay, USAR, during the inaugural Joint WPS Academic Forum Workshop hosted by the U.S. Army War College (USAWC) in August 2020. Dr. Lauren Mackenzie of MCU volunteered to chair the selection committee, which included Lieutenant Colonel Dana Perkins, USAR, from USAWC, Associate Professor Brenda Oppermann from the Naval War College, and Dr. Susan Yoshihara from the Defense Security Cooperation University—all of whom generously gave of their time and knowledge to review the papers submitted for consideration. Command and General Staff College student Major Sarah Salvo's, paper titled "The Effect of Hegemonic Masculinities on the Endemic of Sexual Misconduct in the United States Army" was selected as the winner, and she was recognized with the USAWC WPS coin as well as a certificate and letter from Air Force brigadier general Rebecca Sonkiss, deputy director of Counter Threats and International Cooperation (J-5), the Joint Staff, Washington, DC.

As of this writing, PME schools from around the United States are selecting their writing award recipients, and it has been the goal of this monograph not only to showcase the "best of the best" across branches of Service and levels of learning but also to pave the way for future recognition of military students' research and writing devoted to the relevance of WPS at tactical, operational, and strategic levels. As an additional incentive, the Academic Year 2022 "Best of" WPS Writing Award winner will receive a $500 cash prize donated by Colonel Oswald-Hrutkay in her retirement.

A Word of Thanks

It will come as no surprise to those familiar with military culture that many of the uniformed leaders of the WPS effort in recent years are transitioning out of their roles in 2022. A huge debt of gratitude is owed to Colonel Oswald-Hrutkay and Lieutenant

Colonel Perkins for their leadership roles in the Joint WPS Academic Forum in recent years, and we welcome Lieutenant Colonel Amanda Clare, USAR, as she takes over the reigns as the new forum leader. Since the release of the DOD *WPS Strategic Framework and Implementation Plan*, a variety of new working groups have emerged across the DOD, and the Joint WPS Academic Forum exists to bring together military and civilian leaders dedicated to integrating WPS into their respective PME institutions and training programs. The forum has benefited tremendously from the WPS initiatives undertaken by Colonel Oswald-Hrutkay and Lieutenant Colonel Perkins to advance the work of military students and educators. The WPS academic community—with its various educational initiatives spanning the PME spectrum—have grown and strengthened as a result of their dedication.

Lauren Mackenzie, PhD
Professor of Military Cross-Cultural Competence
Marine Corps University
Quantico, VA

APPENDICES

Appendix A
Setting the Example

International security institutions play a critical role in the advancement of Women, Peace, and Security (WPS) principles. The North Atlantic Treaty Organization (NATO) was not expected to become a stakeholder in the implementation of the United Nations Security Council Resolution (UNSCR) 1325, yet NATO saw how the WPS agenda applied to the broader discussion of security as it related to alliance efforts.[1] Today, NATO is considered a leader in the implementation of the WPS agenda. NATO is integrating a gender perspective across its three core tasks: collective defense, crisis management, and cooperative security.[2] As both an international organization and a security alliance, NATO has significant impact and influence on behavior with its member states and with host nation states where NATO forces operate.

Prior to the passage of UNSCR 1325, NATO leaders advocated for awareness and recognition about women's experience in the armed forces and to increase their status through participation.[3] In 1954, NATO addressed the roles of female servicemembers during the Annual Review Committee on National Military Service. NATO established the Committee on Women in NATO force in 1976.[4] In the 1990s, NATO forces witnessed gender-based sexual violence as a tactic in the Balkans. Today, NATO advocates for combating conflict-related sexual violence with the international community. As a result, NATO has military guidelines on prevention of and response to conflict-related sexual violence, with formal policy in development.[5] Further, NATO recognizes that gender dynamics may have security implications. In Afghanistan, NATO implemented relative components of the WPS agenda by incorporating a gender perspective into the entire planning cycle of its Resolute Support Mission.[6] Gender perspectives improved situational awareness and human terrain understanding. NATO saw how Afghan women could participate in the political process and could contribute to their own communities through engagement with deployed servicewomen. Female Engagement Teams were viewed as force multipliers to counterinsurgency operations.[7] Increasing the understanding of how security

---

[1] "NATO and WPS: How an Unlikely Pair Became Inseparable," *NATO WPS Bulletin* 2 (Autumn 2020), 4–7.
[2] Charlotte Isaksson, "Integrating Gender Perspectives at NATO: Two Steps Forward, One Step Back," in *Women and Gender Perspectives in the Military* (Washington, DC: Georgetown University Press, 2019), ed. by Robert Egnell and Mayesha Alam, 225.
[3] Lisa A. Aronsson, *Listen to Women* (Washington, DC: Atlantic Council, Scowcroft Center for Strategy and Security, 2020), 42–45.
[4] "NATO and WPS," 4.
[5] "NATO Stands with the International Community to Address Sexual Violence in Conflict," NATO, 23 April 2021.
[6] Aronsson, *Listen to Women*, 42–45.
[7] Katharine Wright, "NATO's Adoption of UNSCR 1325 on Women, Peace and Security Make the Agenda a Reality," *International Political Science Review* 37, no. 3 (May 2016): 350–61, https://doi.org/10.1177/0192512116638763.

challenges affect men, women, boys, and girls differently is at the center of NATO's gender approach to missions and operations.[8] NATO's support of WPS underscores the value of operationalizing gender perspective to improve military effectiveness.

To support WPS objectives, NATO initially adopted two guiding documents. The first guidance, *NATO/Euro-Atlantic Partnership Council (EAPC) Women, Peace, and Security: Policy and Action Plan*, was published in 2007. It is NATO's guide to WPS implementation on three principles: inclusion, integration, and integrity.[9] The second document is NATO's *Bi-Strategic Command Directive 040-001*, which outlines the military implementation of WPS. The directive created the gender advisory structure to facilitate gender perspective implementation into all military structures and activities—from doctrine to an integral element of the planning cycle for missions and operations.[10]

Additionally, NATO has put WPS principles into practice by integrating a gender perspective in all its activities to support an inclusive organization. As such, NATO has policies and plans that advance a gender perspective throughout both its political and military structures. Moreover, incorporating a gender perspective is evolving how NATO works across multiple lines of effort. Some examples include conducting gender analysis of early warning indicators, examining the gender dimensions of both victims and actors of violent extremism, and cooperation with women's civil society groups for potential impact on NATO policy and practice.[11]

From the NATO viewpoint, inclusiveness is not just a numbers game—i.e., more women in the armed forces—but meaningful participation and change. Key lines of effort include organizational changes that include the use of gender-inclusive language, gender balance in both civilian and military structures, and encouraging member states and mission partners to do the same. By setting the example to model gender-inclusive practices, NATO demonstrates its WPS commitment to the alliance's 30 member states and 19 partner nations.[12] NATO's robust WPS policies and strategies are essential to help influence the behavior of member states who are responsible for WPS implementation through their respective national action plans (NAPs). At times, putting policy into action is a challenge when 7 of the NATO member states do not have a WPS NAP.[13] Despite challenges to WPS implementation across all member states and partner nations, NATO is an international standard setter.

Integrity to uphold WPS principles is a fundamental enabler to lead by example with partner organizations who also promote gender-related issues such as the UN, the European Union, the African Union, and the Organization for Security and Cooperation in Europe.[14]

---

[8] "Women are Vital to a Stronger NATO," NATO, 8 March 2021.
[9] Aronsson, *Listen to Women*, 42–45.
[10] "Women, Peace and Security," NATO, 14 May 2021.
[11] "Women, Peace and Security."
[12] "Women, Peace and Security."
[13] Aronsson, *Listen to Women*, 42–45.
[14] Aronsson, *Listen to Women*, 44.

NATO sets the example through leadership. NATO's commitment to WPS is maintained by male and female civilian and military leaders throughout NATO's organizations. From gender advisors and gender focal points across all levels—strategic, operational, and tactical—to the high-level focal point secretary general's special representative for Women, Peace and Security.[15] NATO also maintains the Committee on Gender Perspective that promotes gender integration as an integral element of design and implementation of policies, programs, and military operations.[16] NATO's current special representative for WPS, Clare Hutchinson, states that cultural change across an institution or organization requires more than assigning a team to oversee gender equality.[17] Successfully incorporating a gender perspective will depend on the level of integration and recognition of WPS principles beyond the required mandate.

NATO senior leaders routinely meet to review progress on WPS implementation. For instance, in dedicated sessions leaders work together to identity challenges and opportunities to incorporate gender into NATO's work on resilience and countering violent extremism. For the first time, in October 2020, NATO defense ministers discussed how NATO has furthered the WPS agenda and what needs to be done to overcome the challenges to progress.[18] NATO leadership recognizes that implementing the WPS agenda requires practical aspects to work collectively to ensure accountability. Through policy and practice, NATO is committed to the advancement of gender equality in the global security architecture. NATO Secretary General Jens Stoltenberg asserts that WPS principles have a significant contribution to make in NATO's adaption to a complex and challenging security environment.[19]

Therefore, NATO's future goals in WPS include evolving a gender perspective to include emerging security threats like climate change, pandemics, cyber security, and disinformation campaigns.[20] Though progress has been made in implementing WPS principles, NATO WPS leaders recognize there is continued work to be done. Last year NATO established the Leadership Task Force on WPS chaired by the deputy secretary general. The task force is comprised of senior leaders from the International Staff, NATO agencies, and International Military Staff to provide a forum to work across the NATO command structure.[21] To successfully integrate gender perspective requires leadership across the organization.

Successful implementation of WPS also requires understanding and knowledge on how gender principles apply to your area of responsibility. Major General Michele

---

[15] "Women, Peace and Security."
[16] LtCol Rachel Grimes, "Exclusive Interview with UNSCR 1325 as She Turns 19," *NATO Review*, 31 October 2019.
[17] Clare Hutchinson, "Are We There Yet? Implementing the Women, Peace and Security Agenda: If Not Now, When," *NATO WPS Bulletin* 2 (Autumn 2020), 3.
[18] "Are We There Yet?," 2.
[19] "NATO and WPS," 6.
[20] "NATO and WPS," 7.
[21] "NATO and WPS," 6.

Risi, commander to NATO's Mission in Kosovo, supports that gender perspective is a force multiplier and tool to better achieve his mission's mandate. He stresses that WPS should be integrated into all staff procedures and assessment—it is not an add-on capability. Major General Risi also reinforces the value of supportive leadership to the implementation of WPS in NATO missions. He expects his gender advisors to be supported by commanders at headquarters and regional command levels. Based on his experience, he affirms that incorporating a gender perspective to improve operational effectiveness starts with the Commander all the way down the chain of command.[22]

Conflict prevention and resolution are multidimensional and complex. The WPS agenda underpins the necessity of incorporating a gender perspective into peace processes and military operations. NATO has opened the aperture on gender through policy and in practice. Viewing operations and activities through a gender lens broadens the opportunities for innovation and solutions in complex security environments. Stability and prosperity are not possible if half of the population does not participate in decisionmaking and conflict resolution. Senior leaders have a responsibility to lead and champion WPS principles through their respective organizations. Continuous culture change must be advocated by the senior civilian and military leadership to accelerate transformative change. And knowledge, training, and dissemination of WPS guidance must be vertical and horizontal across the organization. NATO sets the example by actively addressing challenges of implementing WPS, exploring how to adapt their approach to the evolving security environment, and seeking ways to strengthen NATO's future role in WPS.

---

[22] "Views from the Field: Interviews With Major-General Jennie Carignan, Commander NATO Mission Iraq, and Major-General Michele Risi, Commander NATO's Mission in Kosovo," *NATO WPS Bulletin* 2 (Autumn 2020), 13–15.

Appendix B
Advising Project Interview

Date/time: _____ Location: _____
Subject #: _____ Interview length: _____

Informed consent: Yes – *continue* / No – *end interview*
Place of birth/current resident: _____
Duty station/work station: _____
Education level: _____
Occupation: _____
Years of advising experience: _____
Positions/jobs in advising: _____
Location of advising work: HQRS / Kabul / TAAC / Other: _____

Detailed Questions:
1. What is your job specialty/MOS and rank?
2. How long have you been in the military/civilian sector?
3. Have you deployed before; where?
   a. Have you been to Afghanistan; how many times?
4. How many days' notice did you get prior to this deployment?
5. When did you arrive to your post?
6. Did you receive a hand over turn over? Please describe.
7. What was your source of deployment (JMD, YIS, MODA, CEW, CTR, SFAB, AFPAK, TDY, etc.)?
8. What was your original billet assignment?
9. Are you still serving in your original billet assignment?
10. How long are you deployed to Afghanistan?
11. Are you an advisor by billet?
12. Do you speak Dari or Pashto?
13. Who and where do you advise? What level do you advise at (TAAC, Ministerial)?
14. How often do you advise per week?
15. What are some of your advising responsibilities?
16. What are limitations of your advising efforts?
17. Is your advisee receptive to your advising suggestions?
18. How long did it take for you and your advisee to connect and build a relationship?
19. Is there anything that could have made the duration of building a relationship shorter in time?

20. Did you receive advisory predeployment training prior to deployment? (MODA, JFTC, RSTE, KLT, APFAK, Country Specific Training)
    a. On a scale of 1–10 (10 being the best, 1 being the least how well did your training prepare you to be an advisor?
    b. On a scale of 1–10 (10 being the best, 1 being the least) how was the quality of your predeployment advisory training?
    c. What improvements could be implemented to enhance the quality of your predeployment advisor training?
21. Did you read any publications on advising prior to deployment? (RS SFA 4.0 Guide, Advising MS TTPs for Adv. Foreign Sec. Forces, CALL Senior Level Advising, etc.)
22. Did you read any publications on Afghanistan prior to deployment? (Books, articles, CALL Newsletter Afghanistan Culture Sep 10, etc.)
23. Did you attend in-country Resolute Support Orientation Training?
    a. On a scale of 1–10 (10 being the best, 1 being the least) how well did the RSO training prepare you for your advisor duties?
24. Did you attend in-country Military Advising Group Orientation sposored by CSTC-A?
    a. On a scale of 1–10 (10 being the best, 1 being the least) how well did the MAG-O training prepare you for your advisory duties?
25. What improvements could be implemented to enhance the quality of your in-country advisor training?
26. Please describe gaps in predeployment or in-country advisor training that could improve the quality of your advising duties.
27. In your opinion, are there any advisor skills that should be taught to advisors? Please describe.
28. Have you read the RS OPLAN?
29. Please describe the Resolute Support Mission.
30. Please describe the term *functional-based security force assistance*.
31. Please describe the difference between *security force assistance* and *defense institution building*.
32. What problems have you experienced while advising or in the RSM in general?
33. What recommendations do you have for future improvement in advising in Afghanistan?
34. If you could tell the senior general officers about anything in the performance of your advising duties, what would you tell them?
35. Any last comments you wish to share?

Interview notes:

___

Appropriate for follow-up interview? Y / N
Describe:

___

Appendix C
Integrating WPS into PME

It is recommended that U.S. Army professional military education (PME) institutions develop a phased approach to incorporating Women, Peace, and Security (WPS) into curricula. First, at precommissioning and primary levels, lessons should focus on building a foundational knowledge of WPS with an introduction to policy and strategy. Additionally, primary-level instruction should focus on integrating and operationalizing gender at the tactical level. Second, at the intermediate level, education should build on WPS foundational knowledge and provide further instruction on conducting a gender analysis. In addition to the lesson plan below, if time allows, a scenario-driven exercise where students provide a gender analysis is recommended.

Gender should be incorporated into operational-level planning. Last, senior level PME should focus on integrating WPS principles into strategic-level planning.

Intermediate Level PME Sample Lesson Plan
Lesson Title: Gender and Conflict

> Adding a gender perspective has the potential to transform the traditional military paradigm by including and creating an increased understanding of the importance of non-traditional security issues.
> ~Robert Egnell[23]

Introduction

Women and peace have been associated throughout history largely based on the assumption that women are more emotionally empathetic and inherently nonviolent. More modern research has moved beyond women as symbols and actors of peace and has sought to establish the role of gender perspectives in peace and security. One particular study found that the participation of women in all aspects of peace negotiations led to a 20 percent increase in the probability of the peace agreement lasting longer than two years and a 35 percent increase in the probability of a peace agreement lasting 15 years.[24]

In 2001, predicated on the role of women in conflict prevention and resolution, the United Nations Security Council passed Resolution (UNSCR) 1325 on Women, Peace and Security (WPS), which called for the full participation of women in peace and security initiatives. UNSCR 1325 and the eight subsequent resolutions provide a framework for the implementation of gender perspective in the pursuit of international security. The UN resolutions include a range of complex, multilayered

---

[23] Robert Egnell, "Gender Perspectives and Military Effectiveness: Implementing UNSCR 1325 and the National Action Plan on Women, Peace and Security," *Prism* 6, no. 1 (March 2016).
[24] Marie O'Reilly, Andrea Ó Súilleabháin, and Thania Paffenholz, *Reimagining Peacemaking: Women's Role in Peace Process* (New York: International Peace Institute, 2015), 41–42.

issues such as inclusion of gender in all facets of peacekeeping operations, the representation of women in conflict resolution, gender perspectives mainstreaming, training reformations, and the recognition and protection of women, girls, and boys from conflict-related threats. Most of all, the resolution calls for gender equality and urges the international community to take the necessary steps to put the plan into action.

In 2011, President Barrack Obama signed Executive Order 13,595 establishing the *The United States National Action Plan on Women, Peace, and Security*.[25] In 2017, the Women, Peace, and Security Act of 2017 was signed into law, strengthening efforts for the meaningful participation of women in conflict prevention and peace building. The law ensures congressional oversight of how the United States promotes and implements women's meaningful participation in conflict prevention and resolution. Furthermore, the United States released its national strategy on WPS in 2019, outlining four primary lines of effort.

1. Seek and support the preparation and meaningful participation of women in conflict-related decision-making.
2. Promote the safety and protection of women's and girls' human rights.
3. Adjust U.S. international programs to improve outcomes in equality for, and the empowerment of, women.
4. Encourage partner nations to adopt policies to improve the meaningful participation of women.[26]

To achieve the goals outlined across the four lines of effort given in the national strategy, the Department of Defense released the *Women, Peace, and Security Strategic Framework and Implementation Plan* (WPS SFIP) in June 2020. The plan organizes WPS implementation along three defense objectives that include modeling and employing WPS, promoting partner-nation women's participation, and promoting the protection of partner-nation civilians.[27]

In connecting gender and operational effectiveness, scholars note that considering gender can lead to increased credibility and security and an increase in information-gathering capability. The enhanced information gathering obviously leads to the increase of force protection, providing information that can lead to the findings of weapons, explosive devices, or high value targets.[28] However, beyond increased force protection, scholars have recognized that the situation and security of women in a country is often the best indicator of how likely that country is to be involved in

---

[25] *The United States National Action Plan on Women, Peace, and Security* (Washington, DC: Office of the President, White House, 2016).
[26] *United States Strategy on Women on Women, Peace, and Security* (Washington DC: Office of the President, White House, 2019), 6.
[27] *Women, Peace, and Security Strategic Framework and Implementation Plan* (Washington, DC: U.S. Department of Defense, 2020).
[28] Sahana Dharmapuri, "Just Add Women and Stir," *Parameters* 41, no. 1 (Spring 2011): 57.

conflict.[29] Empirical results lead to the conclusion that human security (namely, the security of women) is linked to national and international security. Better security leads to a more stable state and decreased likelihood of harboring terrorists, violating human rights, and requiring intervention from the international community. U.S. national security is dependent upon "stable, prosperous, and democratic societies abroad."[30] Women's participation, gender inclusion, and gender perspectives are necessary to maintain stable societies and increase force protection in the armed forces.

Student Learning Objectives
1. Understand how gender can shape the way we examine conflict.
2. Apply a gender analysis to conflict prevention and resolution.
3. Gain an appreciation for how gender perspectives can affect military effectiveness.

Required Reading (pages: 53)
Chinkin, Christine, Mary Kaldor, and Punam Yadav. "Gender and New Wars." *Stability International Journal of Security and Development* 9, no.1 (2020): 1–13. http://doi.org/10.5334/sta.733. (10 pages)

Egnell, Robert. "Gender Perspectives and Military Effectiveness." *Prism* 6, no 1 (March 2016): 73–87. (14 pages)

Anderlini, Sanam Naraghi. "Mainstreaming Gender in Conflict Analysis: Issues and Recommendations." *Social Development Papers: Conflict Prevention and Reconstruction* no 33 (2006): 1–9. (9 pages)

MacKenzie, Megan. "Securitization and Desecuritization: Female Soldiers and the Reconstruction of Women in Post-Conflict Sierra Leone." *Security Studies* 18, no. 2 (2009): 241–61. https://doi.org/10.1080/09636410902900061 (20 pages)

Supplemental Reading (pages: 15, plus 4:07 viewing minutes)
Gender and Conflict Analysis Toolkit for Peacebuilders.
London, UK: Conciliation Resources, 2015. Introduction, chapter 2. (15 pages)

Saferworld. "Gender Analysis of Conflict: Why is it important?," 5 July 2016, video, YouTube, 4:07.

---

[29] Valerie M. Hudson, "What Sex Means for World Peace," *Foreign Policy*, 24 April 2012.
[30] *Women on the Frontlines of Peace and Security* (Washington DC: National Defense University Press, 2014), 22.

Issues for Discussion

1. What is the significance of considering gender when studying conflict?
2. What roles may masculinity and femininity play in conflict?
3. How may military operations change or become more effective when gender is considered?
4. How could the Southern Confederacy have gained the support of Britain during the U.S. Civil War? Or were Union diplomatic actions too powerful?
5. How can a gender analysis affect conflict resolution and prevention?

CHRONOLOGY

Women, Peace, and Security: How Did We Get Here?

2020 — DOD Women, Peace, and Security Strategic Framework and Implementation Plan
Provides DOD objectives to accomplish WPS strategy and act; fulfills WPS Act and strategy requirment to establish plan.

2019 — U.S. Strategy on WPS
Four-year whole-of-government strategy to guide U.S. government WPS implementation; satisfies Executive Order 13,595 requirements; tasks DOD, DOS, USAID, and DHS with creating implementation plans.

2017 — U.S. Women, Peace, and Security (WPS) Act
United States enacts the first comprehensive law in the world on WPS; tasks DOD, DOS, USAID, and DHS as implementing departments.

2011 — Executive Order 13,595
Mandates WPS implementation; United States becomes the 50th country with a national action plan on WPS that translates UNSCR 1325 into national context.

2000 — UNSCR 1325
First global recognition of the disproportionate impact of conflict on women and girls and their necessary role in preventing and resolving it.

Source: *Women, Peace, and Security Strategic Framework and Implementation Plan* (Washington, DC: U.S. Department of Defense, 2020).

GLOSSARY OF KEY WPS CONCEPTS AND TERMS

Empowerment is the process of gaining authority, power, confidence, and control to perform various acts or duties autonomously.[31]

Equality is the right of different groups of people to have a similar social position and receive the same treatment.[32]

Equity is a situation in which everyone is treated fairly according to their needs and no group of people is given special treatment.[33]

Gender refers to the socioculturally constructed roles, rights, privileges, and responsibilities of men, women, boys, and girls.[34]

Gender analysis is an examination of the relationships and interactions between men and women, their access to and control of resources, the constraints they face relative to each other, and sociocultural power structures.[35]

Gender balance refers to efforts that ensure both men and women have equal opportunities and access to matters in the instiotions of society (namely: religion, economy, education, culture, and polity). Notably, when applied to the military, gender balance does not equate to the ensuring equal numbers of men and women within units.[36]

Gender-based violence is an umbrella term for any harmful threat or act directed at an individual or group based on actual or perceived biological sex, gender identity, and/or expression, sexual orientation, and/or lack of adherence to varying socially constructed norms around masculinity and femininity. It is rooted in structural gender inequalities, patriarchy, and power imbalances. Gender based violence is typically characterized by the use or threat of physical, psychological, sexual, economic, legal, political, social, and other forms of control and/or abuse. Gender based violence impacts individuals across the life course and has direct and indirect costs to families, communities, economies, global public health, and development.[37]

Gender integration refers to the strategies applied in policy/program assessment, design, implementation, and evaluation to take gender norms into account and to compensate for gender-based inequalities.

---

[31] *Cambridge Dictionary*, s.v. "empowerment," accessed 10 May 2022.
[32] *Cambridge Dictionary*, s.v. "equality," accessed 10 May 2022.
[33] *Cambridge Dictionary*, s.v. "equity," accessed 10 May 2022.
[34] DOD Introduction to Women, Peace, and Security course and DOD Women, Peace, and Security Implementation course, both Joint Knowledge Online, accessed 10 May 2022, hereafter JKO WPS training courses.
[35] JKO WPS training courses.
[36] JKO WPS training courses.
[37] JKO WPS training courses.

- For DOD, *gender integration* often refers to integrating women into the military.
- Outside of DOD, *gender integration* can refer to "the integration of gender perspectives," particularly among U.S. interagency partners and civil society organizations.[38]

Gender mainstreaming, according to NATO, is "a strategy for making the concerns and experiences of women and men an integral dimension of the design, implementation, monitoring, and evaluation of policies, programs, and military operations."[39] According to the European Institute for Gender Equality, it "involves the integration of a gender perspective into the preparation, design, implementation, monitoring, and evaluation of policies, regulatory measures, and spending programmes with a view to promoting equality between women and men, and combating discrimination."[40]

Gender perspective exposes gender differences and how being treated as a man or woman in society shapes a person's needs, interests, and security.[41]

Sex refers to the biological and physiological differences between males and females, such as the anatomy of an individual's reproductive system and genetic differences.[42]

Sex disaggregated data is data on individuals collected and tabulated separately for women and men.[43]

UNSCR 1325 is the United Nations Security Council Resolution on Women, Peace Security adopted unanimously by the UN Security Council on 31 October 2000. It acknowledged the disproportionate and unique impact of armed conflict on women and girls and recognizes the critical role that women play in peace processes.

United States Strategy on Women, Peace, and Security was published in 2019 in response to the Women, Peace, and Security Act of 2017. The strategy seeks to increase women's meaningful participation in political and civic life by ensuring women are empowered to lead and contribute, equipped with the necessary skills to support and

---

[38] JKO WPS training courses.
[39] *NATOTerm*, s.v. "gender mainstreaming," accessed 10 May 2022.
[40] "What is Gender Mainstreaming?," European Institute for Gender Equality, accessed 10 May 2022.
[41] JKO WPS training courses.
[42] JKO WPS training courses.
[43] "Sex-disaggregated Data," European Institute for Gender Equality, accessed 10 May 2022.

succeed, and supported to participate through access to opportunities and resources.[44] Women's contributions as agents of change in conflict prevention and resolution increase peace-building capacity and long-term maintenance of peace.

Women, Peace, and Security (WPS) agenda is a particular theme of the UN Security Council's work under its responsibility for the maintenance of international peace and security that the council has considered at formal meetings. UNSCR 1325 and nine other resolutions are the outcomes of the UN Security Council's WPS agenda.

---

[44] *United States Strategy on Women, Peace, and Security* (Washington, DC: Office of the President, White House, 2019), 2.

SELECTED BIBLIOGRAPHY

Baseline Soldier Physical Readiness Requirements Study. Iowa City: University of Iowa Virtual Soldier Research Program 2020.

Bridges, Donna and Debbie Horsfall. "Increasing Operational Effectiveness in UN Peacekeeping Toward a Gender Balanced Force." *Armed Forces and Society* 36, no. 1 (October 2009): 125. https://doiorg/10.1177/0095327X08327818.

Carter, Ash. "No Exceptions: The Decision to Open All Military Positions to Women." BelferCenter.org, December 2018.

Concepts and Definitions: Women, Peace and Security in NATO. Brussels, Belgium: North Atlantic Treaty Organization, 2019.

Connell, R. W. and James Messerschmidt. "Hegemonic Masculinity: Rethinking the Concept." *Gender and Society* 19, no. 6 (December 2005): 832. https://doi.org/10.1177/0891243205278639.

Cook, Joana. *A Woman's Place: US Counterterrorism since 9/11.* New York: Oxford University Press, 2020.

Defense Advisory Committee on Women in the Services 2019 Annual Report. Alexandria, VA: Defense Advisory Committee on Women in the Services, 2019.

Defense Advisory Committee on Women in the Services 2020 Annual Report. Washington, DC: Department of Defense Advisory Committee on Women in the Services, 2021.

Department of Defense Annual Report on Sexual Assault in the Military, Fiscal Year 2019. Washington DC: U.S. Department of Defense, 2020.

Department of Defense Board on Diversity and Inclusion Report: Recommendations to Improve Racial and Ethnic Diversity and Inclusion in the U.S. Military. Washington, DC: U.S. Department of Defense, 2020.

Developing Today's Joint Officers for Tomorrow's Ways of War: The Joint Chiefs of Staff Vision and Guidance for Professional Military Education and Talent Management. Washington, DC: Office of the Joint Chiefs of Staff, 2020.

Diaz, Pablo Castillo, Simon Tordjman, Samina Anwar, Hanny Queta Beteta, Colleen Russo, Ana Lukatela, and Stephanie Ziebell. *Women's Participation in Peace Negotiations: Connections between Presence and Influence,* 2d ed. New York: UN Women, 2012.

Egnell, Robert. "Gender Perspectives and Military Effectiveness: Implementing UNSCR 1325 and the National Action Plan on Women, Peace and Security." *Prism* 6, no. 1 (March 2016): 72–89.

Egnell, Robert and Mayesha Alam, eds. *Women and Gender Perspectives in the Military: An International Comparison.* Washington, DC: Georgetown University Press, 2019.

Enloe, Cynthia. *Bananas, Beaches and Bases: Making Feminist Sense of International Politics.* Berkeley: University of California Press, 2014.

Esper, Mark T. "Immediate Actions to Address Diversity, Inclusion, and Equal Opportunity in the Military Services." DOD memo. 14 July 2020.

Esper, Mark T. "Message to the Force on DOD Diversity and Inclusiveness." Speech, U.S. Department of Defense, Washington, DC, 18 June 2020.

Eys, Mark and Albert Carron. "Role Ambiguity, Task Cohesion, and Task Self-Efficacy." *Small Group Research* 32, no. 3 (June 2001): 356–73. https://doi.org/10.1177/104649640103200305.

Executive Order 13,595—Instituting a National Action Plan on Women, Peace, and Security. Washington, DC: The White House, 2011.

Gaska, Frank, Ryan Voneida, and Ken Goedecke. "Unique Capabilities of Women in Special Operations Forces." *Special Operations Journal* 1, no. 2 (November 2015): 105–11. https://doi.org/10.1080/23296151.2015.1070613.

Gender Equality: Women's Rights in Review 25 Years after Beijing. New York: UN Women, 2020.

"Guiding Documents," UN Women (website). Accessed 11 January 2021.

Hardison, Chaitra, Susan Hosek, and Chloe Bird. *A Review of Best Practice Methods*, vol. 1 of *Establishing Gender-Neutral Physical Standards for Ground Combat Occupations*. Santa Monica, CA: Rand, 2018.

Hudson, Valerie M., Bonnie Ballif-Spanvill, Mary Caprioli, and Chad F. Emmett. "The Heart of the Matter: The Security of Women, the Security of States." *Military Review* (May–June 2017): 18–34.

Hudson, Valerie M., Donna Lee Bowen, and Perpetua Lynne Nielsen. *The First Political Order: How Sex Shapes Governance and National Security Worldwide.* New York: Columbia University Press, 2020.

Hutchinson, Susan. "Leading the Operationalization of WPS." *Security Challenges* 14, no. 2 (2018): 124–43.

Inclusion, Justice, Security: Women Peace and Security Index 2019/20. Washington, DC: Georgetown Institute for Women, Peace and Security, 2020.

Interim National Security Strategic Guidance. Washington, DC: Office of the President, White House, 2021.

Johnson-Freese, Joan. *Women, Peace and Security: An Introduction.* New York: Routledge, 2019.

Just the Facts: A Selected Annotated Bibliography to Support Evidence-Based Policy Making on Women, Peace, and Security. Broomfield, CO: One Earth Future Foundation, 2019.

King, Anthony. "The Female Combat Soldier." *European Journal of International Relations* 22, no. 1 (March 2016): 122–43. https://doi.org/10.1177/1354066115581909.

Kotter, John P. *Leading Change.* Boston: Harvard Business School Press, 2012.

Krause, Jana, Werner Krause, and Piia Branfors. "Women's Participation in Peace Negotiations and the Durability of Peace." *International Interactions* 44, no. 6 (August 2018): 985–1016. https://doi.org/10.1080/03050629.2018.1492386.

Mackenzie, Megan. *Beyond the Band of Brothers: The US Military and the Myth that Women Can't Fight.* New York: Cambridge University Press, 2015.

Marquis, Jefferson P., Coreen Farris, Kimberly Curry Hall, Kristy N. Kamarck, Nelson Lim, Douglas Shontz, Paul S. Steinberg, Robert Stewart, Thomas E. Trail, and Jennie W. Wenger. *Improving Oversight and Coordination of Department of Defense Programs that Address Problematic Behaviors among Military Personnel: Final Report.* Santa Monica, CA: Rand, 2017.

Moore, Emma. "Women in Combat: Five-Year Status Update." Center for a New American Security, 31 March 2020.

National Security Strategy of the United States of America. Washington, DC: Office of the President, White House, 2017.

NATO/EAPC Women, Peace and Security Policy and Action Plan 2018. Mons, Belgium: NATO/Euro-Atlantic Partnership Council, 2018.

Oppermann, Brenda. "Hawks, Doves, and Canaries: Women in Conflict." *Small Wars Journal*, 13 August 2014.

O'Reilly, Marie and Andrea Ó Súilleabháin. *Women in Conflict Mediation: Why it Matters.* New York: International Peace Institute, 2013.

Oudraat, Chantal de Jonge, Ellen Haring, Diorella Islas, and Ana Laura Velasco. *Enhancing Security: Women's Participation in the Security Forces in Latin America and the Caribbean.* Washington, DC: Women in International Security, 2020.

Preventing Conflict, Transforming Justice, Securing the Peace: A Global Study on the Implementation of United Nations Security Council Resolution 1325. New York: UN Women, 2015.

Raz, Guy. "Why We Can No Longer See Sexual Violence as a Women's Issue." *TED Radio Hour*, National Public Radio. Interview transcript. 1 February 2019.

Reimagining Peacemaking: Women's Roles in Peace Processes. New York: International Peace Institute, 2015.

Report of the Fort Hood Independent Review Committee. Washington, DC: Office of the Secretary, U.S. Army, 2020.

Roberts, Delia, Deborah L. Gebhardt, Steven E. Gaskill, Tanja C. Roy, and Marilyn A. Sharp. "Current Considerations Related to Physiological Differences Between the Sexes and Physical Employment Standards." *Applied Physiology, Nutrition, and Metabolism* 41, no. 6 (June 2016). https://doi.org/10.1139/apnm-2015-0540.

Rohwerder, Brigitte. *Lessons from Female Engagement Teams: GSDRC Helpdesk Research Report 1186.* Birmingham, UK: Governance and Social Development Resource Centre, University of Birmingham, 2015.

Schaefer, Agnes Gereben, et al. *Implications of Integrating Women into the Marine Corps Infantry.* Santa Monica, CA: Rand, 2015.

Sjoberg, Laura. *Gender, War, and Conflict.* Cambridge, United Kingdom: Polity Press, 2014.

Stability, Army Doctrine Publication 3-07. Washington, DC: Department of the Army, 2019.

Summary of the 2018 National Defense Strategy of the United States of America: Sharpening the American Military's Competitive Edge. Washington DC: U.S. Department of Defense, 2018.

Teaching Gender in the Military: A Handbook. Geneva, Switzerland: Geneva Centre for the Democratic Control of Armed Forces and Partnership for Peace Consortium, 2016.

UN Security Council. Resolution 1325, Women, Peace and Security, S/RES/1325. 31 October 2000.

United States National Action Plan on Women, Peace, and Security. Washington, DC: Office of the President, White House, 2011.

United States Strategy on Women, Peace, and Security. Washington, DC: Office of the President, White House, 2019.

White House Briefing Room. "Fact Sheet: President Biden Signs Executive Order Enabling All Qualified Americans to Serve Their Country in Uniform." News release. 25 January 2021.

Webster, Joseph E. "Resisting Change: Toxic Masculinity in the Post Modern United States Armed Forces, (1980s–present)." PhD diss., University of Central Oklahoma, 2019.

William M. (Mac) Thornberry National Defense Authorization Act for Fiscal Year 2021, Pub. L. No. 116-283. 2020.

Witkowsky, Anne A. "Integrating Gender Perspectives within the Department of Defense." *Prism* 6, no. 1 (March 2016): 35.

Woetzel, Jonathan, et al. *The Power of Parity: How Advancing Women's Equality Can Add $12 Trillion to Global Growth.* Shanghai: McKinsey Global Institute, 2015.

Women in the Military: Deployment in the Persian Gulf War. Washington, DC: U.S. General Accounting Office, 1993.

Women, Peace, and Security Act of 2017, Pub. L. No. 115-68. 2017.

Women, Peace and Security National Action Plan Development Toolkit. New York: PeaceWomen of Women's International League for Peace and Freedom, 2013.

Women, Peace, and Security Strategic Framework and Implementation Plan. Washington, DC: U.S. Department of Defense, 2020.

Women's Participation in Peace Negotiations: Connections Between Presence and Influence. New York: United Nations Women, 2012.

Women's Rights are Human Rights. New York and Geneva: Office of the United Nations, High Commissioner for Human Rights, 2014.

Woolley, Anita Williams, et al. "Evidence for a Collective Intelligence Factor in the Performance of Human Groups." *Science* 330 (October 2010): 686–88.

INDEX

Aberdeen Proving Grounds (APG), 118
abortion, 87–92; criminalization of, 87–88, 90; law, 87, 88, 90–91
agriculture, 108
Afghan National Army, 192, 202–4, 210
Afghan National Defense and Security Forces (ANDSF), 170–71, 185, 187, 188, 191–92, 194–98, 201–2, 205–10, 212
Afghanistan Papers, 175, 203
Afghan Peace Process, 170
al-Qaeda, 30, 34, 177, 182, 188, 191
artifacts, 124, 127, 134, 142, 144, 145
Army Combat Fitness Test (ACFT), 63–64, 66, 68–69, 71, 73–76, 82
Army People Strategy, 20–21
Army Physical Fitness Test (APFT), 64, 68–69, 75
Army Ranger School Ranger Assessment Phase (RAP), 63; RAP week, 66, 69, 71–74
Army Strategic Implementation Plan, 215, 237, 240
Army Values, 129–31, 144–49, 156–57, 161–63
attitudes of, 5, 20, 53, 63–64, 66–74, 78, 80–81, 123, 132–33, 145, 149, 154–55, 158, 160–63, 178
Australian Defence Force, 215, 222, 229–31

band of brothers, 61–62, 65–66, 70
The Basic School (TBS), 76
Beijing Declaration and Platform for Action: Beijin +5 Political Declaration and Outcome, 98
Berger, CMC Gen David H., 62
bias, 5, 61, 67–72, 86–87, 132, 140, 144, 161, 164; implicit, 164; predictive, 67–68; systemic, 161, 164; unconscious, 144, 164

boys club, 14, 153–55, 159–60
building partner capacity, 176, 180–88, 209, 231

capability gap, 5, 9
capacity building, xii, 177–78, 183, 186, 210, 227
climate, 14, 18, 21, 22, 123–25, 127, 135, 137, 140–41, 144–50, 154–63; command, 14, 22, 124, 140–41, 149, 150; cultural, 31; hostile, 131; organizational, 127, 132–33, 137
climate change, 107–13, 244
cohesion, 18, 63, 64–66, 68–76, 80, 146; group, 65, 131; military, 65; social, 64–65, 69; task, 64, 77, 82; unit, 18, 63, 64–69, 72–73, 77, 131–32, 144, 145
conflict prevention initiatives, 10
conflict-related sexual violence, 242
combat arms, 61–70, 72–81, 223; combat arms occupations, xv, 63
Combat Endurance Test (CET), 64, 82
Combat Exclusion Policy, 61, 63
Combat Fitness Test (CFT), 71, 78
combating trafficking in persons, 227, 233
Comprehensive Special Law for a Life Free of Violence for Women (2010, El Salvador), 90
conflict, ix, xv, xvi, 5, 9, 10, 12, 24–27, 29, 34, 39, 40, 43, 45, 54, 64, 87, 97–100, 119, 120, 164, 169–70, 177–82, 184–87, 189–92, 197–98, 202–4, 206, 210, 215–18, 220, 221, 227, 231, 237, 239; negotiations of, 170, 181, 184; resolution of, xv, 5, 9, 24–26, 40, 93–94, 98, 100, 170, 181, 184, 186, 189, 197, 204, 215, 221, 233–34, 239

countering violent extremist organizations (CVEO), 34, 36
culture, xv–xvi, 5–9, 11, 12, 14, 16, 18–22, 26, 30, 34, 41, 43–45, 52–53, 71, 79–80, 117, 122–64, 169, 176, 180–81, 193, 195, 205, 210; formal, 123, 127, 139, 145, 150, 151, 154, 156; informal, 139, 142, 143, 145, 147, 157; masculine, 133, 137; military, 6, 12, 131, 136, 137, 143, 237, 241
cultural support teams (CSTs), 31–32, 34, 37, 185, 206, 219, 226

decision making, xv, 10, 56, 113, 170, 197, 211, 232
Defense Advisory Committee on Women in the Services (DACOWITS), 13, 223
Defense Objective 1, 12, 13, 20, 42, 189, 222–24
Defense Objective 2, 189, 224–27
Defense Objective 3, 189, 227–28
deforestation, 111
Department of Defense Annual Report on Sexual Assault (2019), 17
Department of Defense Directive 6495.01, Sexual Assault Prevention and Response (SAPR) Program, 16
discrimination, 13–20, 87, 88, 91, 132, 135, 146–48, 161
diversity, xii, 6–18, 21, 26, 36, 53, 56, 68, 74, 138, 146, 164, 165, 172, 194, 205, 219, 220, 237
doctrine, organization, training, materiel, leadership and education, personnel, facilities, and policy (DOTMLPF-P), xii, xvi
domestic violence, 86–87, 179
due process, 88, 92
Dunwoody, Gen Ann, 102

El Salvador, 85–94, 218

empowerment, 28, 45, 94, 98, 113, 179, 189, 210, 221, 234
Equality, Equity, and Eradication of Discrimination Against Women (2011, El Salvador), 91
equity, xv, 6, 11–14, 16, 91, 237
Esper, Mark, 13, 81
espoused beliefs/values, 70, 125, 127, 142, 144, 145, 149–51, 156
ethics, 22, 129
ethical alignment, 130
Executive Order 13,595—Instituting a National Action Plan on Women, Peace, and Security, 10, 25, 39, 99, 221

female engagement team (FET), 6, 26, 29–37, 206, 219, 224
femicide, 85, 87, 91
femininity, 11, 133, 136, 217
Fernandes, Sgt Elder, 122–23
Flynn, Lt Jeannie, 99. *See also* Leavitt, MajGen Jeannie
forced labor, 110
Fort Hood Independent Review Committee (FHIRC), 6, 15, 19, 20, 21, 22, 123, 139–41, 145–64
Fort Hood Prostitution Ring Scandal, 120
Fort Hood, Texas, 6, 14–20, 120–24, 139–41, 144–64

geographic combatant command (GCC), 7, 33, 36, 39, 41–55
gender, ix, xv, xvi, 5–7; gender advisor (GENAD), ix, 34, 36, 42, 46, 51, 52, 55, 92, 101, 172, 185, 197, 204, 208–9, 222–28, 235; gender analysis, 34, 36, 92, 113, 176, 190, 194, 197, 222, 230, 234, 235; gender-based violence, xv, xvi, 6, 9, 11, 16, 18–21, 85–87, 90–94, 98–99,

107, 109, 112, 184, 191, 219, 225, 227, 233; gendered behaviors, 170; gender blindness, 27, 204; gender discrimination, 17, 88, 135, 146–48, 161; gender equality, 39, 45, 53, 90, 98, 195, 197, 205, 210, 218, 226, 233; gender focal point (GFP), 51, 92, 223, 224, 230, 232; gender hostilities, 132, 146–47; gender identity, 11, 70; gender inclusion, 32, 146, 218; gender integration, 10, 62, 146–49, 161, 202, 224; gendered intelligence analysis, 206; gender mainstreaming, 92, 197, 205, 229–32, 237, 240; gender neutral, 14, 27, 61–82; gender-neutral physical fitness test, 61, 72–75, 80; gender-neutral physical standards, 63–67, 72, 78, 80, 82; gender neutrality, 27, 63–64, 67–71, 73, 79, 80; gender norms, 33, 86, 92, 170; gender norming, 68; gender/-ed perspective(s), ix, 5–10, 15, 18, 22–39, 41–56, 90–94, 170–76, 184–98, 202–10, 215–16, 218, 223–33, 236–37, 239; gender-sensitive, 170, 176, 178, 181, 185–87, 198, 201, 203, 208–10; gender studies, 217

Guillén, Spc Jessica, 6, 14, 18–20, 122–23, 141, 152, 164

hegemonic masculinity/-ies, 117, 123–25, 127, 133–49, 152–56, 158–63, 240

Hudson, Valerie, 180, 217–18

human security, xv, 26, 28, 33–34, 218, 223

human smuggling/-ers, 107, 109–10, 113

human trafficking/-ers, 107, 109–13, 227

hypermasculine, 131–33

inclusion, xii, 6, 7, 9, 11–18, 21, 26, 30, 32, 46, 48, 54, 56, 61, 63–64, 68–73, 77, 81, 100, 112, 129, 138, 146, 161, 164, 176–77, 185–86, 189, 192, 202, 204, 206, 211, 218–19, 224, 226, 237, 239

Independent Review Commission on Sexual Assault in the Military, x

inequality/-ies, xv, 7, 39, 41, 64, 93, 98, 133, 136, 180, 184, 190; structural, 11, 85, 94

ingroup, 70

institutional responsiveness, 86

integration, xi, xv, 10, 33, 41, 61–68, 70–82, 86, 90, 125, 146–49, 161, 164, 197, 202, 220, 223–24, 227, 234, 237

Inter-American Court of Human Rights, 89

Interim National Security Strategic Guidance (2021), 20, 37

ISIS-Khorasan (ISIS-K), 192

Islamic State of Iraq and Syria (ISIS), 26–27, 33–34, 170; Islamic State, 26–27, 170

Israeli Defense Forces, 66

Joint Knowledge Online, ix, 224, 236

Joint Women, Peace, and Security Academic Forum, xi

Karpinski, BGen Janis, 101

Kotter, John, xvi, 7, 39, 42–53; *Leading Change*, 7, 43; Leading Change model, xvi, 39, 53

Leavitt, MajGen Jeannie, 99, 102. See also Flynn, Lt Jeannie

Lioness Teams, 29, 31. See also Team Lioness

Lynch, Pvt Jessica, 100, 101

MacArthur, Gen Douglas, 143–44
machismo, 86, 94
machista, 88
Manuela v El Salvador, 89–90
Marine Corps Infantry Officers Course (IOC), 63–64, 66, 69, 71, 76–78
masculine/-ity/-ies, xv, 11, 93, 117, 123, 125, 133–37, 140, 142–45, 149, 152–55, 161, 217, 23; complicit, 134, 153, 159; marginalized, 134; subordinate, 134, 149, 152
Mattis, Gen James, 62, 80
Mekong delta, 108–9, 112
McKinney, SMA Gene C., 118
McQueen, SGC Gregory, 120
McRaven, Adm William, 31, 37, 102
migration, 107–13; climate-induced migration, 112; internal migration, 109; labor migration, 110
migrant(s), 107, 109–12
miscarriage, 88–89
motherhood, 13, 88
Multisystem Ethical Culture Framework, 130, 139
murder, 14, 17–19, 87, 122–23, 141, 152

national action plan (NAP), 10, 25, 39, 85, 90–93, 99–101, 110, 169, 176, 185, 189, 191, 198, 202, 207–10, 221, 225–29
national defense strategy (NDS), 11–12, 45, 183, 190; *National Defense Strategy of the United States of America* (2018), 11
national security strategy (NSS), 10–12, 20, 27, 100, 183; *National Security Strategy of the United States of America* (2017), 11, 183
Nelson, Katherine, 127, 127, 130, 139, 142, 157

North Atlantic Treaty Organization (NATO), 27, 30–31, 33, 35–36, 48, 171, 175, 183–85, 191, 194–99, 203–10, 223
noncommissioned officers (NCOs), 23, 56, 72, 130, 148, 151–53, 159, 163
nongovernmental organization, 35, 112

O'Donnell, Col Mark, 32
Olson, Adm Eric, 31, 37
organizational culture, xvi, 20, 22, 39, 41, 43–44, 52–53, 63, 68, 70, 72, 76, 80–81, 117, 123–27, 129–47, 153, 156–57, 161–62; informal, 117, 130, 139
outgroup, 70

patriarchy/-al, 5, 7, 11, 134, 152, 159, 179, 181, 210
peace building, ix, 24, 25, 34, 90, 100, 181, 182, 184, 186, 190, 204, 205, 215, 221
peacekeeping, ix, 5, 9, 10, 25, 26, 33, 35, 99, 178, 217–18, 224
Peacekeeping and Stability Operations Institute (PKSOI), 224
Petraeus, Gen David H., 101
physical abilities, 61
physical fitness, xv, 61–82
Physical Fitness Test (PFT), 61, 62, 63, 64, 69, 71–79
post-conflict, xv, 10, 26, 40, 170, 177, 178, 181, 182, 184, 186
professional military education (PME), ix–xii, xv–xvi, 22–23, 36, 56–57, 97, 216, 222, 224, 228, 232, 240–41

Ranger Physical Fitness Test, 63

Ranger School, 6, 48, 63, 73–75, 79, 82
rape, 87, 88, 90, 117, 125, 144, 184; rape culture, 117
real men, 135–36, 143
Report of the Fort Hood Independent Review Committee, 6, 124, 139, 141
reproductive rights, 87, 92
Resolute Support Mission (RSM), 170–76, 185, 190–97, 199–211
Resolute Support Operational Plan (RS OPLAN), 205
Robinson, Gen Lori, 102
Robinson, Spc Aaron, 122

security force advising, 169
security force assistance, 182, 191–92, 201, 203, 205, 234, 236
sex/-ual, 11, 14, 18, 26, 64, 74, 79, 92, 112, 118, 120, 143, 148, 164, 169, 172, 180, 187, 190, 195, 217, 219, 227, 231; sex trafficking, 110, 227; sexual assault, x, 6, 11, 12, 16–22, 36, 50–52, 87, 100, 117–27, 132, 138–41, 144–56, 158–65; Sexual Assault Prevention and Response (SAPR) Program, 16, 119, 162; sexual exploitation, 19, 110; sexual harassment, x, 6, 9, 11, 12, 16–22, 52, 117–27, 129, 131–33, 137–41, 144–65,. 204; sexual misconduct, xvi, 9, 117, 123, 151, 154, 156, 158, 210, 231, 233, 240; sexual political order, 180; sexual violence, xv–xvi, 22, 90, 98, 135, 154, 156, 158, 184, 191, 225, 227, 233; U.S. Army Sexual Harassment/Assault Response and Prevention (SHARP) program, 17, 18, 119–20, 123, 129, 140, 144–59, 162–63, 233
Schein, Edgar, 127–29, 139, 142, 147

Schuurman, Marriet, 35
Sharia law, 179, 188
Special Operations Forces (SOF), 28, 31–32, 81
stillbirth, 89
strategic framework and implementation program (SFIP), xi, 5, 9, 10, 12, 100, 171, 215, 221, 241
structural inequality/-ies, 85, 94
systemic violence, 19, 85, 91, 119
systemic discrimination, 14, 19, 20, 87, 88

Taliban, 30, 170, 174, 175, 177–82, 188–93, 197–98, 202, 205–7, 211–12, 226
Team Lioness, 101. *See also* lioness teams
trafficking victims, 107, 109–112
train, advise, assist (TAA), 191, 192, 196, 203
Treviño, Linda, 127, 129–30, 139, 142, 157

underlying assumptions, 126, 127–29, 142, 145, 162
uniform standards, 14
United Nations (UN), 5, 9, 24, 25, 27, 28, 39, 42, 110, 184, 191, 205, 215, 218, 222, 229, 239; UN Department of Peacekeeping Operations (DPKO), 218, 222; UN Security Council Resolution (UNSCR), 42; UNSCR 1325, Women, Peace, and Security, 5, 9–10, 24–25, 28, 35–36, 38, 39, 42, 90, 97–99, 102, 169–71, 175–77, 184–86, 189, 191–93, 195–98, 202–3, 205–10, 215, 230, 233, 236
United States National Action Plan on Women, Peace, and Security (WPS NAP), 10, 39, 90–93, 99–100, 189, 221, 226, 228, 250

United States Strategy on Women, Peace, and Security (2019), xi, 10, 25, 40, 171, 185, 189
U.S. Africa Command (USAFRICOM), 227
U.S. Agency for International Development (USAID), 10, 212
U.S. Air Force Air War College (AWC), xv, 97, 101–2
U.S. Army, xi–xii, xv–xvi, 6, 13, 14, 15, 18, 19, 21, 29, 31, 32, 33, 44, 47, 48, 50, 52, 63, 68, 72–80, 82, 98, 117–24, 127, 129–33, 136, 138–52, 155–65, 196, 201, 215–16, 222, 224–29, 232–35, 237, 240
U.S. Army Command and General Staff College, xv–xvi, 127, 240
U.S. European Command (USEUCOM), 33
U.S. Indo-Pacific Command (USINDOPACOM), 33–34
U.S. Marine Corps, x, xii, 29, 62, 63, 64, 66, 75, 76–80, 82, 222, 224, 240
U.S. Navy, 136, 219
U.S. Southern Command (USSOUTHCOM), 226

Verveer, Melanne, 42
Vietnam, 27, 107–13
village stability operation (VSO), 31–32
violence, xv–xvi, 6, 9, 11–12, 16, 18–21, 24, 26–27, 54, 85–94, 98–99, 107, 109, 112, 133–37, 144, 160, 174, 175, 177, 179, 211, 218, 227, 233, 234, 237; against women, xv, 85–88, 91, 93; endorsement of, 85, 91; legitimization of, 85, 91; multisided, 86

white nationalist/-ism, 24, 136

William M. (Mac) Thornberry National Defense Authorization Act for Fiscal Year 2021, 5
Women, Peace, and Security (WPS): Women, Peace, and Security Act of 2017, 10, 25, 40, 100, 185, 190, 204, 209, 211, 221, 226, 228, 233; WPS agenda, xv, 5, 6, 9, 24–26, 36–37, 54, 56, 57, 184, 189, 197, 205, 210, 215, 230–31, 232–34, 239; WPS strategy, xi, 37, 97, 185, 204, 209, 234, 237, 239
Women, Peace, and Security Strategic Framework and Implementation Plan (2020) (WPS SFIP), xi, 5, 8, 9, 10–12, 16–18, 25, 36, 40, 42–47, 54, 100–2, 171, 189, 197, 209, 211, 215, 221–22, 225–29, 234
women's rights, 5, 15, 47, 85–87, 91–92, 178–79, 184, 188, 198, 211

ABOUT THE EDITORS

Dr. Lauren Mackenzie is Marine Corps University's Professor of Military Cross-Cultural Competence. She also chairs the Marine Corps University faculty council and serves as an adjunct professor of military/emergency medicine at the Uniformed Services University of Health Sciences in Bethesda, Maryland. She earned her master of arts and doctorate in communication from the University of Massachusetts Amherst and has taught intercultural and interpersonal communication courses throughout the DOD for the past 12 years. Dr. Mackenzie leads the MCU Women, Peace & Security Scholars Program, and her research is devoted to the impact of cultural differences on difficult conversations. She has written a range of articles and book chapters pertaining to end-of-life communication, relationship repair strategies, and, most recently, the role of failure in education. She is coeditor with Dr. Kerry B. Fosher of the 2021 book *Rise and Decline of U.S. Military Culture Programs: 2004–20* (MCU Press) and is the 2020 civilian recipient of the Marine Corps University Rose Award for Teaching Excellence.
Her email address is: lauren.mackenzie@usmcu.edu.

Lieutenant Colonel Dana Perkins is a U.S. Army Reserve 71A/microbiologist serving as the director for Women, Peace, and Security Studies with the Office of the Provost, U.S. Army War College, in an Individual Mobilization Augmentee position. She has a doctorate in pharmacology and experimental therapeutics from the University of Maryland in Baltimore and she was awarded the D1 (Countering Weapons of Mass Destruction Advisor) and D7 (Strategic Studies) Additional Skill Identifiers, as well as a D7A (Defense Support to Civilian Authorities) Personnel Development Skill Identifier. She is a graduate of JS J5 Operationalizing WPS L100 and L200 courses for gender focal points and gender advisors, respectively. In her civilian capacity, Dr. Perkins serves as a senior science advisor with the Office of the Assistant Secretary for Preparedness and Response, U.S. Department of Health and Human Services.
Her email address is dana.s.perkins.mil@army.mil.